AF443565

Student Study Guide to Accompany

# INTRODUCTION TO CHEMISTRY

**Martha J. Gilleland**
California State College,
Bakersfield

Prepared by

**Rebecca Williams**
Richland College
Dallas, Texas

**West Publishing Company**
St. Paul     New York
Los Angeles     San Francisco

COPYRIGHT © 1986 by WEST PUBLISHING CO.
50 West Kellogg Boulevard
P.O. Box 64526
St. Paul, MN 55164-9979

All rights reserved
Printed in the United States of America

ISBN  0-314-96991-8

# CONTENTS

# PREFACE

This study guide accompanies the text, INTRODUCTION TO CHEMISTRY, by Martha Gilleland.  For each chapter in the text, the study guide contains the following sections

> SUMMARY OUTLINE
> SOLUTIONS TO SELECTED TEXT QUESTIONS AND PROBLEMS
>     and PRACTICE PROBLEMS
> SELF-TEST
> ANSWERS TO PRACTICE PROBLEMS
> ANSWERS TO SELF-TEST

The SUMMARY OUTLINE lists the headings for each section of the text followed by statements of key concepts.  Fundamental relationships, rules, and guidelines are summarized using formulas, "maps", tables, and listings.

The second section, SOLUTIONS TO SELECTED TEXT QUESTIONS AND PROBLEMS ..., utilizes a format established in the text; an example problem with a detailed solution is followed by a practice problem.  For the study guide example problems, I provided detailed solutions to a number of end-of-chapter text problems.  (In most cases, these problems have answers in Appendix C in the text.)  The solutions provide additional explanation, strategies for solving the problem, and a thorough treatment of the mathematics involved.  Following these examples, similar problems are given for practice.  Students can utilize the problem-solving strategies, assess understanding, and generalize concepts while solving the practice problems. Solutions to the practice problems are given at the end of each chapter in the study guide.

The Solutions/Practice Problem section of the study guide
can be very helpful, if it is used correctly.  I recommend that
you read the text first, working examples and exercises as you
read.  Circle the number of the end-of-chapter exercises in the
text that have detailed solutions in the study guide.  As you
work these problems, check your answers in Appendix C.  Then,
refer to the solution in the study guide.  You can affirm the
method that you used to solve the problem or get additional
explanation in order to make corrections.  Once you understand
the method, work the practice problem and check your answer.
Notice the similarities and differences in the example problem
and the practice problem.

Calculator setups are introduced in the text's Math Tips and
reinforced throughout the solutions shown in the study guide.
I've shown many answers as they appear on the calculator,
rounding them to the correct number of significant figures for
the final answer.  This was done to call attention to the need
for analysis of the calculator answer.

The SELF-TEST section contains ten multiple-choice questions.
These questions are typical questions and problems covering the
text material.  The Self-Test provides further opportunities for
self-assesment and is suitable as a closure activity for the
chapter.  The final two sections of the study guide give solutions to the practice problems and the answers to the self-test
questions.

I would like to acknowledge and thank Kathy Simpson, who
expertly typed and corrected the manuscript; my colleague at
Richland College, Jo Blackburn, and Ralph Blackburn, who care-
fully proof-read the manuscript and checked the problems; and
Mike, Neal, and Kara Williams, who provided encouragement all
along the way.  Above all, though, this work is for my mother,
Venita Grundon, who gave so much time and energy in support of
this project.

Rebecca Williams
Dallas, Texas
November, 1985

# CHEMISTRY: ORIGINS AND SCOPE

CHAPTER<br>**1**

---

SUMMARY OUTLINE

---

**1.1 The Origins of Chemistry**

Chemistry had its beginnings in the ancient art of alchemy.

Alchemists' two major goals were to prolong life and to convert common, inexpensive metals into gold for use in the elixir of life.

**1.2 The Nature of Chemistry**

Chemistry is the study of matter and its transformations.

**1.3 Studying Chemistry**

Attend all class meetings.

Read assigned material before the lecture.

Take good notes.

Ask questions of yourself and your instructor.

Study every day.

**1.4 States of Matter**

Matter is anything that occupies space and has mass.

The physical states of matter are solid, liquid and gas.

When matter changes from one physical state to another, the change is called a change of state.  A change of state does not change the identity of the substance.

## 1.5    Classification of Matter

A phase is a state of matter having clearly defined and distinguishable boundaries.

Homogeneous matter has the same properties and composition throughout.  It consists of only one phase.

Heterogeneous matter has variable composition and properties throughout.  It is a mixture of two or more phases.

## 1.6    Subclassification of Matter

Heterogeneous matter can be subclassified as a mixture of different substances in separate phases or as a mixture of different phases of the same substance.

Homogeneous matter can be subclassified as a homogeneous mixture of different substances in one phase or as a pure substance.  A pure substance has a definite and constant composition.

## 1.7    Physical and Chemical Changes

A physical change does not alter the identity of a substance.

A chemical change alters the identity of a substance.

## 1.8    Properties of Matter

Identifying features and characteristics of matter are called properties.

Physical properties can be determined by observation and measurement of a substance without changing its identity.

Chemical properties can be determined by observation and measurement of a substance as it undergoes a chemical change.

## 1.9    Conservation of Mass in Chemical Reactions

The law of conservation of mass states that matter is neither created nor destroyed during a chemical change.

## 1.10    Energy

Energy is defined as the capacity to do work.  Work is done when an object is moved a distance.

All energy can be classified as kinetic or potential energy.

Kinetic energy is possessed by a moving object.  Potential energy is energy stored in an object due to its position, condition, or composition.

Energy can exist in many forms: heat, mechanical, electrical,
chemical and light energy.  Energy can be changed from one
form to another.  However, the law of conservation of energy
states that energy cannot be created or destroyed.

### 1.11   Energy Changes in Chemical Reactions

Energy changes accompany all chemical reactions and many
physical changes.

These energy changes take many different forms, including the
release of heat energy (an exothermic process) and the
absorption of heat energy (an endothermic process).

### 1.12   Energy and Matter

Einstein's equation is $E = mc^2$, where E = energy, m = mass and
c = speed of light ($3 \times 10^{10}$ cm/sec).

Mass and energy are interchangeable.  Since the square of the
speed of light is a very large number, a small change in the
amount of mass can produce enormous amounts of energy.

Chemical reactions involve negligible changes in mass.  Nuclear
reactions result in significant mass changes, and thus large
energy changes.

Since mass and energy are interchangeable, the two laws of
conservation can be combined into one.  The total mass and
energy of the universe remains constant.

---

### SOLUTIONS TO SELECTED TEXT STUDY QUESTIONS AND PROBLEMS
### and
### PRACTICE PROBLEMS

---

3.   What is the physical state of each of the following?

The state of matter for a particular substance depends on the
temperature and pressure at the time of the observation.
Notice in the examples below that temperature is given.
Pressure is understood to be 1 atmosphere, the average
pressure at sea level.

c.   Water at 500°F

   At the boiling point of a substance, the liquid state
   changes to the gaseous state.  Since water boils at 212°F,
   it will be a gas at 500°F.

d.   Air at 200°F

   At room temperature, 72°F, air exists in the gaseous state,
   so at even higher temperatures it will be a gas.

***Practice Problem A

What is the physical state of each substance at the following temperatures?  Consult the table of melting points and boiling points given below.  (Answer questions in the space provided).

a.   -100°C

helium — *gas*

magnesium — *solid*

mercury — *solid*

b.   Room temperature, 22°C

helium — *gas*

magnesium — *solid*

mercury — *liquid*

c.   400°C

helium — *gas*

magnesium — *solid*

mercury — *gas*

|            | melting point | boiling point |
| ---------- | ------------- | ------------- |
| helium     | -272.2°C      | -268.6°C      |
| magnesium  | 651°C         | 1107°C        |
| mercury    | -38.87°C      | 356.58°C      |

4.      Give a name to each of the following processes.

        Each process below represents a physical change because the
        identity of the substance is not altered.  State changes are
        the most common physical changes.  The state changes are
        melting (solid ⟶ liquid), freezing (liquid ⟶ solid),
        evaporation (liquid ⟶ gas), condensation (gas ⟶ liquid or
        gas ⟶ solid), and sublimation (solid ⟶ gas).

    b.  Water is changed to ice.

        It is common to refer to liquid water as just water.  The
        change implied here is liquid water ⟶ solid water, which
        is called freezing.

    c.  Alcohol spontaneously disappears from an open container.
        Liquid alcohol changes to gaseous alcohol (also called
        alcohol vapor), which is evaporation.

    e.  Water bubbles vigorously when heated.

        A state change occurs from liquid to gas.  However
        evaporation occurs on the surface of the liquid.  This
        process is more correctly identified as boiling.  Boiling
        is a special form of evaporation in which the state
        change from liquid to gas occurs within the body of the
        liquid through bubble formation.

        Consult Appendix C in the text for answers to parts a, d and f.

14.     Classify each of the following as a heterogeneous mixture, a
        homogeneous mixture or a pure substance.

    a.  Alcohol dissolved in water.
        Since the alcohol dissolves in the water, only one phase
        is present.  This phase contains two substances, so
        alcohol in water is a homogeneous mixture.

    b.  Mist
        Mist results from fine droplets of water in the air.  Two
        phases exist containing different substances, so mist is a
        heterogeneous mixture.

    e.  Partially frozen pop.

        Two phases are present, the liquid and solid pop.  A
        mixture of different phases of the same substance is a
        heterogeneous mixture.

        Consult Appendix C in the text for answers to parts c, d and f.

***Practice Problem B

        Classify each of the following as a heterogeneous mixture, a
        homogeneous mixture, or a pure substance.  Indicate how many

phases are present.   (Answer questions in the space provided).

a.   sugar *— pure subst. (one phase)*

*homo.*  b.   water and sugar — *homo. (one phase)*

*hetero*  c.   liquid and solid water *hetero. (two phases)*

*hetero* d.   water and mineral oil *hetero. (two phases)*

*hetero*  e.   water, sugar and mineral oil *hetero. (two phases)*

*hetero*  f.   liquid and solid water and mineral oil *hetero. (three phases)*

18.   Classify each of the following as a physical or chemical change.

No new substances are formed when a physical change occurs. Common physical changes are state changes  and changes in size or shape.   A chemical change results in the formation of one or more new substances.

*Phy.* b.   The crushing of grapes.

Crushing grapes changes the size and shape of the grapes, so a physical change has occurred.

*chem.* d.   The tarnishing of copper.

The new substance formed on the surface of copper when it tarnishes is the result of a chemical change.

*phy.* e.   The expansion of water when it freezes.

A state change occurs when water freezes, which is a physical change.

Consult Appendix C in the text for answers to parts a, c, and f.

22.   Tell whether each of the following statements describes a physical or chemical property.

Physical properties are observable without changing the substance.   Examples of physical properties are color, odor, size, shape, physical state, density, taste, melting point and boiling point.   Chemical properties are observable only when a substance undergoes a chemical change.   Chemical properties are often described by reference to other substances.

*phy.* a.   Sugar is a white solid at room temperature.

Two physical properties are mentioned, color and physical state.

b.  Copper conducts electricity.

Electric current passing through a copper wire will not produce any new substances.  Electrical conductivity is a physical property.

c.  Metals are corroded by acid.

The word corroded means to wear away.  The metal chemically combines with the acid to form new substances.  The "disappearance" of the metal is one evidence of the chemical change.  Another is the formation of hydrogen gas as bubbles on the surface of the metal.  Since a chemical change occurs, metals being corroded by acid is a chemical property.

Consult Appendix C in the text for answers to parts d, e, and f.

***Practice Problem C

Classify the following properties of magnesium as physical or chemical properties.  (Answer in the space provided).

a.  burns with a dazzling white flame

b.  melting point, 651°C

c.  silvery white in color

d.  tarnishes slightly in air

e.  one third lighter than aluminum

30.  Classify each of the following processes as exothermic or endothermic.

a.  The boiling of water.

Heat must be added to liquid water before it will boil. The addition of heat in this physical change is an endothermic process.

c.  The formation of ice crystals.

Heat must be removed to change liquid water into ice. The removal of heat is an exothermic process.

Consult Appendix C in the text for answers to parts b and d.

---

## SELF-TEST

Circle the correct answer in the following multiple choice questions.

1.  Which of the following correctly describes the liquid state of matter?

a.   fixed volume, fixed shape
b.   variable volume, variable shape
c.   fixed volume, variable shape
d.   variable volume, fixed shape

2.   The melting point of acetic acid is 16.6°C and the boiling point of acetic acid is 118.5°C.  Which of the following physical states is correct for the temperature given?

a.   liquid, 15.0°C
b.   gas, 101°C
c.   solid, room temperature
d.   liquid, 40°C

3.   Which of the following correctly describes homogeneous mixtures?

a.   one phase, one substance
b.   one phase, two substances
c.   two phases, one substance
d.   two phases, two substances

4.   Acetone has a melting point of −95.4°C and a boiling point of 56.2°C.  Which of the following changes of state is correct for the temperature change given?

a.   −90° to −96°C, freezing
b.   55°  to 57°C, condensation
c.   −96° to −90°C, sublimation
d.   0° to 57°C, melting

5.   Which of the properties of potassium is correctly identified as a physical or chemical property?

a.   reacts rapidly in air; physical property
b.   soft, easily cut with a knife; chemical property
c.   catches fire spontaneously on water; chemical property
d.   produces hydrogen gas when mixed with water; physical property

6.   Which of the following statements is true?

a.   Matter and energy cannot be conserved in all reactions, either matter or energy is lost.
b.   In a chemical change, matter can be conserved, but not energy.
c.   No energy changes can occur in a physical change.
d.   Chemical changes involve relatively small changes in energy when compared to nuclear reactions.

7.   Which of the following changes is correctly identified as exothermic or endothermic?

a.   salt + water = homogeneous mixture + energy; exothermic
b.   energy + liquid mercury = gaseous mercury; exothermic
c.   hydrogen + oxygen = water + energy; endothermic
d.   liquid water = solid water + energy; endothermic

8.  Which of the following statements is not true?

   a.  as a ball rolls uphill, kinetic energy is converted to
       potential energy
   b.  chemical energy cannot be converted to heat energy
   c.  work is done only when an object is moved a distance
   d.  certain substances contain potential energy due to their
       composition

9.  In Einstein's equation, $E = mc^2$,

   a.  c represents the number of calories of heat energy
       liberated when matter is converted to energy.
   b.  a small loss of mass, m, will produce a large quantity of
       energy, E.
   c.  m represents the length in meters that an object is moved
       when work is done.
   d.  mass and energy are shown to be interchangeable, but mass
       increases always result in increases in energy.

10.  Which of the following statements is true?

   a.  heterogeneous matter must be uniform throughout
   b.  heterogeneous matter must contain two different substances
   c.  heterogeneous matter has variable composition
   d.  heterogeneous matter cannot contain three phases

---

### ANSWERS TO PRACTICE PROBLEMS

Practice Problem A

   a.  helium, gas; magnesium, solid; mercury, solid
   b.  helium, gas; magnesium, solid, mercury, liquid
   c,  helium, gas; magnesium, solid, mercury, gas

Practice Problem B

   a.  pure substance; one phase
   b.  homogeneous mixture; one phase
   c.  heterogeneous mixture; two phases
   d.  heterogeneous mixture; two phases since water and mineral
       oil do not mix
   e.  heterogeneous mixture; two phases, one containing sugar
       and water, the other containing mineral oil
   f.  heterogeneous mixture; three phases

Practice Problem C

   a.  chemical property
   b.  physical property
   c.  physical property
   d.  chemical property
   e.  physical property

---

## ANSWERS TO SELF-TEST

1. c
2. d
3. b
4. a
5. c
6. d
7. a
8. b
9. b
10. c

# SCIENTIFIC MEASUREMENTS



---

SUMMARY OUTLINE

**2.1    Precision and Accuracy**

Repeated measurements of the same quantity are precise if the values are very close to each other.

A measurement is accurate if the value is very close to the true value.

**2.2    Significant Figures**

Significant figures are those digits in a measurement that are known with certainty plus one digit that is an estimate.

Numbers that are counted or defined have an infinite number of significant figures.

**Calculations and Significant Figures**

In multiplication and division, the answer must contain the same number of significant figures as the term with the least number of significant figures.

In addition and subtraction, the answer must contain the same number of decimal places as the term with the least number of decimal places.

**2.3    Scientific Notation**

A number written in scientific notation has the general form $N \times 10^{exponent}$, where N is a number between 1 and 10.

2.4  <u>Systems of Measurement</u>

The metric system, based on decimals, is composed of one unit for each type of measurement.  Standard prefixes are used to increase or decrease the size of the unit (Table 2-3).

The SI system specifies the use of certain metric units to standardize the metric system (Table 2-2).

2.5  <u>Conversion of Units</u>

A given unit can be converted to a required unit by the multiplication of a fraction, arranged to cancel the given unit.

$$\left( \frac{\cancel{\text{given unit}}}{} \right) \left( \frac{\text{required unit}}{\cancel{\text{given unit}}} \right)$$

When numbers are written in the fraction so that the numerator equals the denominator, the fraction is equal to one.  The fraction is called a unit factor or conversion factor.

2.6  <u>Mass and Weight</u>

Mass is a measure of the amount of matter in an object.

Weight is a measure of the force exerted on the object's mass by the pull of gravity.

2.7  <u>Density</u>

Density is the amount of mass in a unit volume.

Density is a characteristic property since its value does not depend on sample size.

$$\text{Density} = \frac{\text{mass}}{\text{volume}} = \frac{m}{v}$$

Common units of density are $\frac{g}{mL}$ or $\frac{g}{cm^3}$ .

2.8  <u>Specific Gravity</u>

The density of a substance relative to that of water at 4°C is specific gravity.

$$\text{specific gravity} = \frac{\text{density of substance in g/mL}}{\text{density of } H_2O \text{ at } 4°C \text{ in g/mL}}$$

Specific gravity has no units.

2.9  <u>Units of Energy</u>

The calorie (cal) is a metric unit once defined as the quantity of energy that will raise the temperature of 1 g of $H_2O$ from 14.5°C to 15.5°C.

The calorie has been redefined in terms of the SI unit of energy, the joule (J).

$$1 \text{ cal} = 4.184 \text{ J}$$

## 2.10    Temperature

Heat is a form of energy.

Temperature indicates the coldness or warmth of an object.

Temperature can be measured on the Fahrenheit, Celsius or Kelvin scale.

Equations for converting temperatures

$$°F = 1.8°C + 32$$
$$K = °C + 273$$

## 2.11    Specific Heat

Specific heat is a measure of the amount of heat needed to raise the temperature of 1 g of a substance 1°C.

$$\text{specific heat} = \frac{\text{cal}}{g \cdot \Delta T}$$

where $\Delta T$ is the temperature in °C.

Specific heat has units $\dfrac{\text{cal}}{g \cdot °C}$ .

---

SOLUTIONS TO SELECTED TEXT STUDY QUESTIONS AND PROBLEMS
and
PRACTICE PROBLEMS

---

13.    Express each of the following numbers in scientific notation.

a.    732.6

Writing a number in the form $N \times 10^{\text{exponent}}$ requires 3 steps.

1)    For the value of N to be between 1 and 10, move the decimal behind the first integer.

$$732.6 \longrightarrow 7.326$$

2)    The numerical value of the exponent is determined by the number of places the decimal was moved.

$$7.326 \times 10^{2}$$

3)    The sign of the exponent is determined by the direction the decimal was moved.  Moving the decimal to the left requires a positive exponent.  It is not necessary to memorize this fact.  Since

$$10^{+\text{exponent}} > 1, \text{ multiplication by } 10^{+\text{exponent}}$$
$$\text{and, increase the size of N}$$

$$10^{-\text{exponent}} < 1, \text{ multiplication by } 10^{-\text{exponent}} \text{ will}$$
$$\text{decrease the size of N}$$

When 732.6 is rewritten as 7.326, to correctly reflect the larger size of the original number, multiplication by $10^{\text{exponent}}$ must make 7.326 a larger number.  The exponent must be positive.

$$7.326 \text{ x } 10^2$$

b.   100.4

Move the decimal 2 places to the left and use the exponent 2.  Since 1.004 is much less than 100.4, multiplication by $10^{\text{exponent}}$ must make 1.004 a larger number.  The exponent must be positive.

$$1.004 \text{ x } 10^2$$

c.   7,000,000

Move the decimal 6 places to the left and use the exponent 6.  Since 7.000000 is less than 7,000,000 the exponent must be positive.

$$7.000000 \text{ x } 10^6$$

Writing measurements in scientific notation may allow for some of the zeroes to be dropped in the number above.  In the number 7,000,000 it is assumed that the zeroes are written to show the position of the decimal point.  In scientific notation the exponent shows the position of the decimal point.  Writing the zeroes after the decimal indicates a greater precision than is implied by the original number.  The answer should be written

$$7 \text{ x } 10^6$$

d.   0.0538

Move the decimal 2 places to the right and use the exponent 2.  Since 5.38 is greater than the original number, multiplication by $10^{\text{exponent}}$ must make 5.38 a smaller number. The exponent must be negative.

$$5.38 \text{ x } 10^{-2}$$

Consult Appendix C in the text for answers to part e and f.

***Practice Problem A

Express each of the following numbers in scientific notation. (Work problems in the space provided).

a.   1,375

b.   0.00305

c.   962

d.   6.54        *6.5 x 10*

15.   Express each of the following as ordinary numbers without the powers of 10.

a.   $5.689 \times 10^2$

The ordinary number will be larger than the number written in scientific notation since multiplication by $10^{+exponent}$ will increase the size of N.  Move the decimal 2 places, making the number larger.

568.9

c.   $1.499 \times 10^{-7}$

The ordinary number will be smaller than the number written in scientific notation since multiplication by $10^{-exponent}$ will decrease the size of N.  Move the decimal 7 places, making the number smaller.

0.0000001499

Notice that zeroes must be added so that the decimal can be moved 7 times.

Consult Appendix C in the text for answers to part b, d, e and f.

***Practice Problem B.

Express each of the following numbers written in scientific notation as ordinary numbers without powers of 10.  (Work problems in the space provided).

a.   $7.39 \times 10^4$        *73,900*

b.   $8.00 \times 10^{-2}$        *.0800*

c.   $4.546 \times 10^{-1}$        *.4546*

d.   $1.19 \times 10^2$        *119.*

19.   Make the indicated conversions.

a.   67.3 nm to m

Converting units using the unit factor method requires 3 steps.

1)   Determine the unit relationship (consult Table 2-2 or 2-3).  Knowledge of the prefixes in Table 2-3 is recommended since the same prefixes are used in all metric to metric conversions.

$$n \text{ means } 10^{-9}$$

$$nm \text{ means } 10^{-9} \text{ m}$$

$$1 \text{ nm} = 10^{-9} \text{ m}$$

2)   Write down the quantity to be converted (given quantity) followed by a fraction that allows cancellation of the unwanted unit and introduction of the required unit.

$$6.73 \text{ nm} \times \frac{? \text{ m}}{? \text{ nm}} =$$

Once the units are arranged, write the numbers in the fraction using the unit relationship.  Since 1 nm = $10^{-9}$ m, then

$$\frac{10^{-9} \text{ m}}{1 \text{ nm}} = 1$$

Multiplication by this unit factor will not change the value of the measurement.

$$67.3 \text{ nm} \times \frac{10^{-9} \text{ m}}{1 \text{ nm}} =$$

Cancel the units

$$67.3 \; \cancel{\text{nm}} \times \frac{10^{-9} \text{ m}}{1 \; \cancel{\text{nm}}} =$$

3)   Complete the calculations.

Rewrite 67.3 in scientific notation, $6.73 \times 10^1$. The problem can now be written

$$6.73 \times 10^1 \; \cancel{\text{nm}} \times \frac{10^{-9} \text{ m}}{1 \; \cancel{\text{nm}}} =$$

When multiplying and dividing numbers written in scientific notation, group the numbers, exponential part and units, then multiply.

$$(6.73)(10^1 \times 10^{-9})(m) =$$

To multiply exponents, remember $(a^m)(a^n) = a^{m+n}$

$$6.73 \times 10^{1+(-9)} \text{ m} = 6.73 \times 10^{-8} \text{ m}$$

b.   84.92 g to kg

1)   Determine the unit relationship.

$$k \text{ means } 10^3$$

$$kg \text{ means } 10^3 \text{ g}$$

$$1 \text{ kg} = 10^3 \text{ g}$$

2)   Write the given quantity, followed by a fraction with units arranged to allow cancellation, and then cancel.

$$84.92 \text{ g} \times \frac{?\ \text{kg}}{?\ \text{g}} =$$

$$84.92 \text{ g} \times \frac{1\ \text{kg}}{10^3\ \text{g}} =$$

3)  Complete the calculations.

$$8.492 \times 10^1 \text{ g} \times \frac{1\ \text{kg}}{10^3\ \text{g}} =$$

Group numbers, exponential part and units.

$$(8.492)\left(\frac{10^1}{10^3}\right)(\text{kg}) =$$

To divide exponents, remember $\dfrac{a^m}{a^n} = a^{m-n}$

$$8.492 \times 10^{1-(3)} \text{ kg} = 8.492 \times 10^{-2} \text{ kg}$$

d.  52.3 mm to cm

1)  When both metric units contain prefixes, the conversion should be done in two steps.  The first
    conversion will be to the basic unit with no prefix.
    The second conversion will be to the required unit.

$$\text{mm} \longrightarrow \text{m} \longrightarrow \text{cm}$$

Two step conversions require 2 unit relationships.

| | | |
|---|---|---|
| m means $10^{-3}$ | | c means $10^{-2}$ |
| mm means $10^{-3}$ m | and | cm means $10^{-2}$ m |
| 1 mm = $10^{-3}$ m | | 1 cm = $10^{-2}$ m |

2)  $$52.3 \text{ mm} \times \frac{?\ \text{m}}{?\ \text{mm}} \times \frac{?\ \text{cm}}{?\ \text{m}} =$$

$$52.3 \text{ mm} \times \frac{10^{-3}\ \text{m}}{1\ \text{mm}} \times \frac{1\ \text{cm}}{10^{-2}\ \text{m}} =$$

3)  $$5.23 \times 10^1 \text{ mm} \times \frac{10^{-3}\ \text{m}}{1\ \text{mm}} \times \frac{1\ \text{cm}}{10^{-2}\ \text{m}} =$$

$$(5.23)\ \frac{(10^1 \times 10^{-3})}{10^{-2}}\ (\text{cm}) =$$

$$(5.23)\left(\frac{10^{-2}}{10^{-2}}\right)(\text{cm}) = 5.23 \text{ cm}$$

For metric to metric conversions, notice that the numbers in
the original measurement are not changed by the conversion.
Since the prefixes in the metric system are powers of 10,
changing from one metric unit to another involves only a
change in the exponent in 10exponent.  The final answer will

have the same number of significant figures as the original measurement.  If the exponent in $10^{\text{exponent}}$ is a whole number, then $10^{\text{exponent}}$ is an exact number.  Multiplying and dividing by metric unit relationships will not alter the number of significant figures.

Consult Appendix C in the text for answers to part c, e and f.

***Practice Problem C

Make the indicated conversions.  (Work problems in the space provided).

a.   435 dL to L

b.   0.663 m to μm

c.   $7.6 \times 10^{-1}$ km to cm

21.   Make the indicated conversions.

a.   14.7  lb to g

As in metric to metric conversions, conversions involving both metric and English units can be solved using the unit factor method.

1)   Determine the unit relationship (consult Table 2-2).

$$1 \text{ lb} = 453.6 \text{ g}$$

2)   Write down the given quantity followed by a fraction that allows cancellation of the unwanted unit and introduction of the required unit.

$$14.7 \text{ lb} \times \frac{? \text{ g}}{? \text{ lb}} =$$

Once the units are arranged, write the numbers in the fraction using the unit relationship.

$$14.7 \text{ lb} \times \frac{453.6 \text{ g}}{1 \text{ lb}} =$$

Cancel the units.

$$14.7 \text{ lb} \times \frac{453.6 \text{ g}}{1 \text{ lb}} =$$

3)   Complete the calculations.

$$14.7 \text{ lb} \times \frac{453.6 \text{ g}}{1 \text{ lb}} = 6667.92 \text{ g}$$
$$= 6.67 \times 10^3 \text{ g}$$

The original data has the fewest number of significant digits (3), so the answer must be rounded off to three significant figures.

Notice that conversions involving both metric and English units result in a change in the numbers and units of the measurement.  A scientific calculator will be used to perform all numerical calculations in the Study Guide.  In most cases, all numbers on the calculator display will be shown.  Then these numbers will be rounded off to the correct number of significant figures to give the final answer.

Unit relationships that relate metric units to English units are generally not exact quantities.  The numbers given in the relationship are usually determined experimentally.  Although most metric to English relationships given in Table 2-2 are known to greater precision than reported, the number of significant figures given in this table must be considered if this is the source used.

d.   13.0 yd to m

   1)   Determine the unit relationship.  Consulting Table 2-2 shows no relationship for yd to m.  The relationship

$$1 \text{ m} = 39.37 \text{ in}$$

can be used along with the known relationship

$$1 \text{ yd} = 36 \text{ in}$$

to complete this conversion in two steps.

$$\text{yd} \longrightarrow \text{in} \longrightarrow \text{m}$$

   2)   Write the given quantity, followed by a fraction with units arranged to allow cancellation, and then cancel.

$$13.0 \text{ yd} \times \frac{? \text{ in}}{? \text{ yd}} \times \frac{? \text{ m}}{? \text{ in}} =$$

$$13.0 \text{ yd} \times \frac{36 \text{ in}}{1 \text{ yd}} \times \frac{1 \text{ m}}{39.37 \text{ in}} =$$

   3)   Complete the calculations.

$$\frac{(13.0)\ (36)}{(39.37)} \ (\text{m}) = 11.887 \text{ m}$$
$$= 11.9 \text{ m}$$

The answer is limited to 3 significant figures from the original data.  The relationship 1 yd = 36 in is exact and the number 36  does not limit significant figures to 2.

There are times, when tables are not available, that it might be helpful to know some metric to English unit relationships. It is impractical and unnecessary to memorize all relation- ships given in Table 2-2.  Memorize only one metric to English relationship for length, mass, and volume.  Select one that is frequently encountered in chemistry.  With this one unit relationship any conversion can be made within that measurement.  For example, problem 21d, worked above, used the relationship 1 m = 39.37 in.  Assume that only the relation-ship 2.54 cm = 1 in is known.  To convert 13.0 yd to m requires the following conversions:

$$yd \longrightarrow in \longrightarrow cm \longrightarrow m$$

$$13.0 \text{ yd} \times \frac{? \text{ in}}{? \text{ yd}} \times \frac{? \text{ cm}}{? \text{ in}} \times \frac{? \text{ m}}{? \text{ cm}} =$$

$$13.0 \text{ yd} \times \frac{36 \text{ in}}{1 \text{ yd}} \times \frac{2.54 \text{ cm}}{1 \text{ in}} \times \frac{10^{-2} \text{ m}}{1 \text{ cm}} = 11.9 \text{ m}$$

Consult Appendix C in the text for answers to parts b, c, e and f.

***Practice Problem D

Make the indicated conversions.  (Work problems in the space provided).

a.   2.55 mg to oz

b.   3.0 in to nm

c.   4.50 qt to mL

36.   A sample of the "heaviest" substance known, osmium, has a mass of 61.88 g and a volume of 2.75 mL.  Calculate the density of osmium.

Substitute the measurements, including units into the formula

$$d = \frac{mass}{volume} = \frac{m}{v}$$

$$= \frac{61.88 \text{ g}}{2.75 \text{ mL}}$$

$$= \left(\frac{61.88}{2.75}\right)\left(\frac{g}{mL}\right)$$

$$= 22.5018 \frac{g}{mL} \quad = 22.5 \frac{g}{mL}$$

Always include units in the solution and the answer to the problem.  The unit of density must contain a mass unit in the numerator and a volume unit in the denominator.  The volume, 2.75 mL limits the answer to 3 significant figures.

38.    Calculate the volume occupied by a 20.0-gram sample of
       mercury, density 13.5939 g/mL.

       Values for density and mass are given, the volume is unknown.
       Rearrange the formula

$$d = \frac{m}{v}$$

       to solve for the unknown, v, then substitute the data into the
       formula.

$$v = \frac{m}{d}$$

$$= \frac{20.0 \text{ g}}{13.5939 \frac{\text{g}}{\text{mL}}}$$

$$= \left(\frac{20.00}{13.5939}\right)\left(\frac{\text{g}}{\frac{\text{g}}{\text{mL}}}\right)$$

$$= 1.4712\left(\text{g} \div \frac{\text{g}}{\text{mL}}\right)$$

$$= 1.4712\left(\cancel{\text{g}} \times \frac{\text{mL}}{\cancel{\text{g}}}\right)$$

$$= 1.47 \text{ mL}$$

       An alternate solution utilizes density as a conversion factor.
       The mass of mercury can be "converted" to volume by writing
       the given quantity followed by a fraction with units arranged
       to allow cancellation of the unwanted unit.

$$20.0 \text{ g} \times \frac{? \text{ mL}}{? \text{ g}} =$$

       The density directly relates mass to volume; 1 mL of mercury
       is equivalent to 13.5939 g of mercury.

$$20.0 \cancel{\text{g}} \times \frac{1. \text{ mL}}{13.5939 \cancel{\text{g}}} =$$

$$\left(\frac{20.0}{13.5939}\right) \; (\text{mL}) = 1.47 \text{ mL}$$

***Practice Problem E

   a.  What mass of silver, density 11.40 g/mL will occupy
       5.00 mL?

   b.  What volume of aluminum, density 2.70 g/mL, weighs
       25.00 g?

43.   Copper has a specific gravity of 8.96.  What volume of copper has a mass of 74.3 g?

$$\text{specific gravity Cu} = \frac{\text{density of Cu}}{\text{density } H_2O \text{ @ } 4°C}$$

In the formula above two quantities are known, the specific gravity of Cu and the density of $H_2O$.  The density of Cu can be calculated.  With this calculated density and the given mass, volume can be found.

$$(\text{specific gravity Cu})\ (\text{density } H_2O) = \left(\frac{\text{density Cu}}{\text{density } H_2O}\right)\left(\text{density } H_2O\right)$$

$$(\text{specific gravity Cu})\ (\text{density } H_2O) = \text{density Cu}$$

$$(8.96)\ (1.0000\ \frac{g}{mL}) =$$

$$(8.96 \times 1.0000)\ \frac{g}{mL} = 8.96\ \frac{g}{mL}$$

Now to find volume, use the density of copper as a conversion factor.

$$74.3\ g \times \frac{?\ mL}{?\ g} =$$

$$74.3\ g \times \frac{1\ mL}{8.96\ g} = 8.29241\ mL$$

$$= 8.29\ mL$$

***Practice Problem F

   a.   The specific gravity of gold is 19.3.  What mass of gold occupies 25.00 mL?

   b.   A piece of pure aluminum weighs 20.65 g and occupies a volume of 7.65 mL.  What is the specific gravity of aluminum?

47.   Make the following temperature conversions.

   c.   37°C to °F

The formula °F = 1.8°C + 32 can be used directly for temperature conversions from °C to °F.

$$°F = 1.8(37) + 32$$
$$°F = 66.6 + 32$$
$$°F = 98.6$$
$$°F = 99$$

The original temperature limits the answer to 2 significant figures.

e.   78°F to °C

To convert from °F to °C, rearrange the equation above to solve for °C.

$$°F - 32 = 1.8°C + 32 - 32$$
$$\frac{°F - 32}{1.8} = \frac{1.8°C}{1.8}$$
$$\frac{°F - 32}{1.8} = °C$$
$$\frac{78 - 32}{1.8} = °C$$
$$\frac{46}{1.8} = °C$$
$$25.6 = °C$$
$$26 = °C$$

The original temperature limits the answer to 2 significant figures.

Consult Appendix C in the text for answers to parts a, b, d and f.

***Practice Problem G

Make the following temperature conversions.  (Work problems in the space provided).

a.   -10°F to °C

b.   -45°C to °F

c.   100°F to °C

51.  How much heat, in calories, is needed to raise the temperature of 100.0 mL of water from 25.0°C to 100.0°C?

Rearrange the formula

$$\text{specific heat} = \frac{\text{cal}}{g \cdot \Delta T}$$

to solve for calories, the unit of heat.

$$(\text{specific heat})(g)(\Delta T) = \frac{\text{cal}}{\cancel{g} \ \cancel{\Delta T}} \ (\cancel{g}) \ (\cancel{\Delta T})$$

$$(\text{specific heat})(g)(\Delta T) = \text{cal}$$

To substitute into this equation, the specific heat of water must be known, $1.00 \ \frac{\text{cal}}{g \cdot {}^\circ C}$. The mass of the water sample is not given. In this case since the volume of water is given, the mass can be calculated. Assume a density of 1.0000 g/mL. Then the mass will be

$$100.0 \ \cancel{mL} \ x \ \frac{1.0000 \ g}{1 \ \cancel{mL}} = 100.0 \ g \ \text{Water}$$

$$\Delta T = T_2 - T_1 = 100.0{}^\circ C - 25.0{}^\circ C$$

Now substitute into the equation above.

$$\left(1.00 \ \frac{\text{cal}}{g \cdot {}^\circ C}\right) (100.0 \ g)(100.0{}^\circ C - 25.0{}^\circ C) = \text{cal}$$

$$\left(1.00 \ \frac{\text{cal}}{\cancel{g} \cdot \cancel{{}^\circ C}}\right) (100.0 \ \cancel{g})(75.0 \ \cancel{{}^\circ C}) = 7{,}500 \ \text{cal}$$

$$= 7.50 \ x \ 10^3 \ \text{cal}$$

The answer is limited to 3 significant figures from the temperature data.

***Practice Problem H

   a. Calculate the specific heat of ethylene glycol if $1.5 \ x \ 10^3$ calories are needed to raise the temperature of 265 g of ethylene glycol from 22.0°C to 32.0°C.

   b. How much heat, in calories, is needed to raise the temperature of 250.0 mL of water from 5.0°C to 22.0°C?

---

SELF-TEST

Circle the correct answer in the following multiple choice questions.

1.  Which series of measurements for the mass of a copper penny
    are the most precise?

    a.   3.065g, 3.145g, 3.199g, 3.099g
    b.   3.122g, 3.145g, 3.107g, 3.099g
    c.   3.004g, 3.287g, 3.149g, 3.301g
    d.   3.265g, 3.028g, 3.306g, 3.155g

2.  How many significant figures in the following measurement?

                        0.040300 g

    a.   2
    b.   3
    c.   4
    d.   5

3.  The correct answer to the following calculation for the
    specific heat of a metal is

    $$\text{specific heat} = \frac{1.00 \times 10^3 \text{ cal}}{99.5 \text{ g} \times (95°C - 62°C)}$$

    a.   0.030 $\dfrac{\text{cal}}{\text{g} \cdot °C}$

    b.   0.305 $\dfrac{\text{cal}}{\text{g} \cdot °C}$

    c.   $3.0 \times 10^{-1}$ $\dfrac{\text{cal}}{\text{g} \cdot °C}$

    d.   $3.05 \times 10^{-1}$ $\dfrac{\text{cal}}{\text{g} \cdot °C}$

4.  565,549 written with 3 significant figures is

    a.   566,000
    b.   $5.65 \times 10^5$
    c.   $5.66 \times 10^5$
    d.   566

5.  The correct SI unit for density is

    a.   g/mL
    b.   $kg/m^3$
    c.   $g/m^3$
    d.   $kg/cm^3$

6.  Which of the following solutions is correct for converting
    7.65 nm to cm?

    a.   7.65 nm $\times \dfrac{1 \text{ m}}{10^{-9} \text{ nm}} \times \dfrac{10^2 \text{ cm}}{1 \text{ m}} =$

    b.   7.65 nm $\times \dfrac{100 \text{ cm}}{1 \text{ nm}} =$

c.  $1 \text{ cm} \times \dfrac{10^{-2} \text{ m}}{1 \text{ cm}} \times \dfrac{7.65 \text{ nm}}{1 \text{ m}} =$

d.  $7.65 \text{ nm} \times \dfrac{10^{-9} \text{ m}}{1 \text{ nm}} \times \dfrac{1 \text{ cm}}{10^{-2} \text{ m}} =$

7.  Calculate the volume of 25.0 g of acetone, the solvent in fingernail polish remover.  The density of acetone is 0.791 g/mL.

    a.  31.6 mL
    b.  19.8 mL
    c.  0.0316 mL
    d.  19.8 g

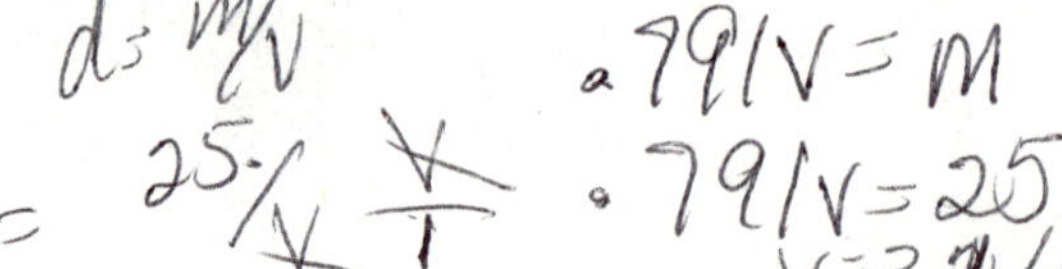

8.  A 100.0 g sample of lead has a volume of 8.85 mL.  What is the specific gravity of lead?

    a.  $8.85 \times 10^{-2}$
    b.  11.3 g/mL
    c.  885
    d.  11.3

9.  The ignition point of an organic substance is 222°C.  What is the temperature in °F?

    a.  454 °F
    b.  432 °F
    c.  104°F
    d.  140°F

10.  Which of the following substances shows the greatest capacity for temperature change with the addition of heat?

    a.  aluminum, specific heat = $0.24 \ \dfrac{\text{cal}}{\text{g} \cdot {}^{\circ}\text{C}}$

    b.  copper, specific heat = $0.093 \ \dfrac{\text{cal}}{\text{g} \cdot {}^{\circ}\text{C}}$

    c.  lead, specific heat = $0.031 \ \dfrac{\text{cal}}{\text{g} \cdot {}^{\circ}\text{C}}$

    d.  sodium, specific heat = $0.29 \ \dfrac{\text{cal}}{\text{g} \cdot {}^{\circ}\text{C}}$

---

## ANSWERS TO PRACTICE PROBLEMS

Practice Problem A

    a.  $1.375 \times 10^{3}$
    b.  $3.05 \times 10^{-3}$
    c.  $9.62 \times 10^{2}$
    d.  $6.54 \times 10^{0}$ or 6.54

Practice Problem B

- a.   73,940
- b.   0.0800
- c.   0.4546
- d.   119

Practice Problem C

a.   $435 \text{ dL} \times \dfrac{10^{-1} \text{ L}}{1 \text{ dL}} =$

$4.35 \times 10^2 \text{ dL} \times \dfrac{10^{-1} \text{ L}}{1 \text{ dL}} =$

$(4.35)\ (10^2 \times 10^{-1})\ \text{L} = 4.35 \times 10^1 \text{ L}$

b.   $0.663 \text{ m} \times \dfrac{1\mu\text{m}}{10^{-6} \text{ m}} =$

$6.63 \times 10^{-1} \text{ m} \times \dfrac{1\mu\text{m}}{10^{-6} \text{ m}} =$

$(6.63)\left(\dfrac{10^{-1}}{10^{-6}}\right)(\mu\text{m}) =$

$(6.63)\,[10^{-1} - (-6)]\ (\mu\text{m}) =$

$(6.63)\,[10^{-1} + (+6)]\ (\mu\text{m}) = 6.63 \times 10^5 \ \mu\text{m}$

c.   $7.6 \times 10^{-1} \text{ km} \times \dfrac{10^3 \text{ m}}{1 \text{ km}} \times \dfrac{1 \text{ cm}}{10^{-2} \text{ m}} =$

$(7.6)\left(\dfrac{10^{-1} \times 10^3}{10^{-2}}\right)(\text{cm}) =$

$(7.6)\,[10^{-1} + (3) - (-2)]\ (\text{cm}) =$

$(7.6)\,[10^{-1} + (3) + (+2)]\ (\text{cm}) = 7.6 \times 10^4 \text{ cm}$

Practice Problem D*

a.   $2.55 \text{ mg} \times \dfrac{10^{-3} \text{ g}}{1 \text{ mg}} \times \dfrac{1 \text{ lb}}{453.6 \text{ g}} \times \dfrac{16 \text{ oz}}{1 \text{ lb}} =$

$\left(\dfrac{2.55 \times 16}{453.6}\right)(10^{-3})(\text{oz}) = 8.995 \times 10^{-5} \text{ oz}$

$= 9.00 \times 10^{-5} \text{ oz}$

The answer is limited to 3 significant figures from the
original data.   1 lb = 16 oz is an exact relationship.

b.   $3.0 \text{ in} \times \dfrac{2.54 \text{ cm}}{1 \text{ in}} \times \dfrac{10^{-2} \text{ m}}{1 \text{ cm}} \times \dfrac{1 \text{ nm}}{10^{-9} \text{ m}} =$

$(3.0 \times 2.54)\left(\dfrac{10^{-2}}{10^{-9}}\right)(\text{nm}) = 7.62 \times 10^7 \text{ nm}$

$= 7.6 \times 10^7 \text{ nm}$

c.   $4.50 \; \cancel{g} \times \dfrac{1 \; \cancel{L}}{1.057 \; \cancel{g}} \times \dfrac{1 \; mL}{10^{-3} \; \cancel{L}} =$

$$\left(\dfrac{4.50}{1.057}\right)\left(\dfrac{1}{10^{-3}}\right) \; (mL) = 4.257 \times 10^3 \; mL$$
$$= 4.26 \times 10^3 \; mL$$

*Solutions may vary slightly depending on choice of conversion factors.

Practice Problem E

a.   $500 \; \cancel{mL} \times \dfrac{11.40 \; g}{1 \; \cancel{mL}} = 57.0 \; g$

b.   $25.00 \; \cancel{g} \times \dfrac{1 \; mL}{2.70 \; \cancel{g}} = 9.26 \; mL$

The density limits the answer to 3 significant factors.

Practice Problem F

a.   The density of gold must be 19.3 g/mL.

$25.00 \; \cancel{mL} \times \dfrac{19.3 \; g}{1 \; \cancel{mL}} = 483 \; g$

b.   Density $= \dfrac{m}{v} = \dfrac{20.65 \; g}{7.65 \; mL} = 2.70 \; g/mL$

Specific gravity $= \dfrac{2.70 \; \cancel{g/mL}}{1.00 \; \cancel{g/mL}} = 2.70$

Practice Problem G

a.   $^\circ C = \dfrac{-10 - 32}{1.8} = \dfrac{-42}{1.8} = -23^\circ C$

b.   $^\circ F = 1.8 \; (-45) + 32$

$= -81 + 32 = -49^\circ F$

c.   $^\circ C = \dfrac{100 - 32}{1.8} = 37.8^\circ C$

Practice Problem H

a.   specific heat $= \dfrac{(1.5 \times 10^3 \; cal)}{(265 \; g)(32.0^\circ C - 22.0^\circ C)}$

$= 5.7 \times 10^{-1} \; \dfrac{cal}{g \cdot {}^\circ C}$

b.   calories $= $ (specific heat)(g)($\Delta T$)

$= \left(1.00 \; \dfrac{cal}{\cancel{g} \cdot \cancel{^\circ C}}\right)(250.0 \; \cancel{g})(22.0 \; \cancel{^\circ C} - 5.0 \; \cancel{^\circ C})$

$= (1.00 \; cal)(250.0)(17.0)$

$= 4.25 \times 10^3 \; cal$

---

## ANSWERS TO SELF-TEST

1. b
2. d
3. c
4. c
5. b
6. d
7. a
8. d
9. b
10. c

# ELEMENTS, ATOMS, AND COMPOUNDS

# CHAPTER 3



---

## SUMMARY OUTLINE

### 3.1 Elements

An element is a pure substance that cannot be separated into simpler substances by ordinary processes.

Hydrogen is the most abundant element in the universe (Table 3-1).

Iron is the most plentiful element on earth (Table 3-2).

### 3.2 The Periodic Table

The periodic table is a graphical arrangement of all chemical elements (Table 3-3). Elements with similar chemical properties are arranged in the same vertical column, called a group or family. Horizontal rows are called periods.

Elements are represented by symbols that can be either one or two letters. The first letter is always upper case, and the second letter is always lower case.

### 3.3 Atoms

An atom is the smallest unit of an element that has all of the properties of the element.

Important concepts of Dalton's atomic theory:

Atoms of one element are distinctly different from atoms of any other element.

Compounds are composed of atoms of two or more elements, present in small, whole-number ratios.
Atoms of the same elements may form more than one compound by combining in different ratios.

Atoms are approximately spherical in shape.  A nucleus takes up a negligible volume in the center of the atom.  Most of the atom is empty space, where electrons move about.

## 3.4 Subatomic Particles

Protons, neutrons and electrons are particles found within atoms.

Protons, neutrons and other particles are found in the nucleus.

Protons, p, have a single positive charge.  Neutrons, n, have no electrical charge.  Electrons, $e^-$, have a single negative charge.

Since atoms have no overall electrical charge, the number of protons equals the number of electrons in an atom.

Atoms can gain or lose one or more electrons to form a charged particle called an ion.

Cations are positively charged ions, formed when atoms lose $e^-$.  When one $e^-$ is lost the cation is $X^+$, if $2e^-$ are lost the cation is $X^{2+}$..., (X is the symbol of the element).

Anions are negatively charged ions, formed when atoms gain $e^-$.  Anions are written $X^-$, $X^{2-}$, $X^{3-}$, ...

## 3.5 Atomic Number

The atomic number is the number of protons in an atom.  All atoms of the same element have the same atomic number.

Elements are arranged in the periodic table in order of increasing atomic number.

## 3.6 Mass Number

Mass number = number of protons + number of neutrons.
Mass number = atomic number + number of neutrons.

## 3.7 Isotopes

Atoms of the same element can have different number of neutrons and thus different mass numbers.  These different atoms of the same element are called isotopes.

Isotopes are represented by the symbol,

$$_A^M E \text{ where, } \begin{array}{l} M = \text{mass number} \\ E = \text{symbol of the element} \\ A = \text{atomic number} \end{array}$$

All elements have isotopes.

### 3.8   Atomic Mass Units

The atomic mass unit, amu, is defined as $\frac{1}{12}$ the mass of the carbon isotope of mass number 12.  According to this definition, $^{12}_{6}C$ has a mass of exactly 12amu.

The approximate masses of the following subatomic particles are

$$\begin{array}{ll} \text{proton,} & \text{1amu} \\ \text{neutron,} & \text{1amu} \\ \text{electron,} & \text{0amu} \end{array}$$

### 3.9   Atomic Weight

The atomic weight of an element is the average of the masses of all isotopes of the element found in nature.

$$\text{Atomic weight} = X_1 m_1 + X_2 m_2 + X_3 m_3 + \ldots$$

where $X_1$ is the fractional abundance of one isotope

$m_1$ is the mass of one isotope

### 3.10   Compounds

A compound is a pure substance composed of two or more elements combined in a definite proportion by weight.

A molecule is a group of two or more atoms held together by the attraction of individual nuclei for electrons in other atoms of the molecule.

Compounds can be composed of ions or molecules.

#### Composition of Compounds

The law of definite proportions states that a given compound always contains the same elements combined in the same proportion by weight.

### 3.11   Chemical Formulas

The formula of a compound contains symbols of all elements in the compound with numerical subscripts indicating the proportions in which the atoms (or ions) exist in the simplest unit of the compound.

When a formula contains symbols in parentheses the atoms in parentheses are bound together as a group.  The subscript following the parentheses refers to the entire group of atoms.

### 3.12   Formula Weights of Compounds

The formula unit of a compound, as described by the formula, is the simplest combination of atoms or ions that has all of the properties of the compound.

Formula weight, FW, is the mass of one formula unit of a compound.  Units of formula weights are amu.

$$FW = n_1W_1 + n_2W_2 + n_3W_3 + \ldots$$

where $n_1$ is the number of atoms of one element in the formula unit.
$W_1$ is the atomic weight of the atom of one element

### 3.13  Percentage Composition

The percentage composition of a compound is the weight percentage of each element in the compound.

$$\%A = \left( \frac{\text{mass of A in } A_xB_y}{\text{FW of } A_xB_y} \right) (100\%)$$

---

SOLUTIONS TO SELECTED TEXT STUDY QUESTIONS AND PROBLEMS
and
PRACTICE PROBLEMS

18.   Each of the following describes the nucleus of an atom of a particular element.  Name the element.

a.  mass number 56; 30 neutrons

In order to identify an element the number of protons, which is the atomic number, must be known.  Since

mass number = number of protons + number of neutrons
56 = number of protons + 30
26 = number of protons

The element is iron, Fe.

Consult Appendix C in the text for answers to parts b through f.

21.   Give the number of protons, neutrons, and electrons in each of the following ions.

a.  $^{137}_{56}Ba^{2+}$

Atoms are neutral particles containing an equal number of protons and electrons.  Ions are charged particles containing a different number of electrons than the parent atom.  Ions and atoms of the same element have the same number of protons.

The barium ion has 56 protons and 81 neutrons (137−56=81). A 2+ charge means that the ion has 2 more positive particles (protons) than negative particles (electrons).  Since the ion must have 56 protons to be barium, 2 fewer electrons are 54.

d.  $^{128}_{52}Te^{2-}$

The tellurium ion has 52 protons and 76 neutrons (128-52=76).
A 2⁻ charge means that the ion has 2 more negative particles
(electrons) than positive particles (protons).  Since the ion
must have 52 protons to be tellurium, 2 more electrons are 54.

Consult Appendix C in the text for answers to parts b, c, e
and f.

***Practice Problem A

Fill in the table below:

| Symbol | Atomic Number | Mass Number | Protons | Neutrons | Electrons |
|---|---|---|---|---|---|
| $^{25}_{13}$Al | 13 | 25 | 13 | 12 | 13 |
|  | 17 | 32 | 17 | 15 | 17 |
| $^{39}_{19}$K+ | 19 | 39 | 19 | 20 | 18 |
|  | 16 | 34 | 16 | 18 | 18 |

22.   Which of the following ions are isoelectronic?

   a.   Br⁻

   b.   $S^{2-}$

   c.   $Sr^{2+}$

   d.   $Li^{+}$

   e.   $Ca^{2+}$

   f.   K+

Isoelectronic ions contain equal number of electrons.  Consult
the periodic table to determine the atomic number and number
of protons of each element.  Using the charge on the ion,
find the number of e⁻.

   a.   Br⁻ 35 protons, 36 electrons (since the ion has a negative
      charge, it must have one more electron than protons.)

   b.   $S^{2-}$ 16 protons, 18 electrons.

   c.   $Sr^{2+}$ 38 protons, 36 electrons (since the ion has a 2+
      charge, it must have 2 fewer electrons than protons.)

d.   $Li^+$ 3 protons, 2 electrons.

e.   $Ca^{2+}$ 20 protons, 18 electrons.

f.   $K^+$ 19 protons, 18 electrons

$Br^-$ and $Sr^{2+}$ have 36 electrons and are therefore isoelectronic. $S^{2-}$, $Ca^{2+}$, $K^+$ are isoelectronic with 18 electrons.

***Practice Problem B

What ion will form for the following elements if each ion is isoelectronic with the argon atom?   (Answer problem in the space provided)

a.   Cl     $Cl^-$, Cl gains 1e⁻ to get 18

b.   Ca     $Ca^{2+}$, Ca loses 2e⁻ to get 18

c.   P      $P^{3-}$, P gains 3e⁻ to 18

d.   K      $K^+$ K loses 1e⁻ to 18

28.   Naturally occurring neon, the gas used in brightly colored electric signs, is composed of 90.92% $^{20}_{10}Ne$, 0.257% $^{21}_{10}Ne$, and 8.82% $^{22}_{10}Ne$.  Calculate the atomic weight of neon.  (Assume a mass of 1.0amu for each neutron and proton).

Assuming a mass of 1.0amu for each neutron and proton, the mass of each isotope is

$^{20}_{10}Ne$, 20.0amu

$^{21}_{10}Ne$, 21.0amu

$^{22}_{10}Ne$, 22.0amu

The fractional abundance of each isotope can be found by dividing the percentage abundance by 100.

$^{20}_{10}Ne$, $\frac{90.92\%}{100} = 0.9092$

$^{21}_{10}Ne$, $\frac{0.257\%}{100} = 0.00257$

$$^{22}_{10}\text{Ne}, \quad \frac{8.82\%}{100} = 0.0882$$

The fractional abundance of an isotope multiplied by its mass, gives the contribution of each isotope toward the average mass.

$$^{20}_{10}\text{Ne} \quad 0.9092 \ \text{X} \ 20.0\,amu = 18.2 \quad amu$$

$$^{21}_{10}\text{Ne} \quad 0.00257 \ \text{X} \ 21.0\,amu = \ 0.0540\,amu$$

$$^{22}_{10}\text{Ne} \quad 0.0882 \ \text{X} \ 22.0\,amu = \ \underline{\ 1.94 \ }\ amu$$

$$20.194 \ amu$$

$$\text{or} \quad 20.2 \quad amu$$

Add each isotope's contribution to find the average mass of neon.  Notice how this procedure follows the general form of

$$\text{atomic weight} = X_1 m_1 + X_2 m_2 + X_3 m_3 + \dots$$

where $X_1$, $X_2$, $X_3$ are fractional abundances of the

isotopes, and $m_1$, $m_2$, $m_3$ are the respective masses of the isotopes.  Each product $(X_1 \cdot m_1)$ gives the contribution of that isotope toward the average mass.

Remember that in addition the number of decimal places in the sum must be the same as the data with the fewest number of decimal places.  In this problem, average mass, 20.2amu, can have only one decimal place.

***Practice Problem C

The naturally occurring isotopes of iron have the following abundances and masses:

$$^{54}_{26}\text{Fe}, \quad 5.84\% \quad 53.9\,amu$$

$$^{54}_{26}\text{Fe} \ .0584 \times 53.9 = 3.15\ amu$$

$$^{56}_{26}\text{Fe}, \quad 91.68\% \quad 56.0\,amu$$

$$^{56}_{26}\text{Fe} \ .9168 \times 56.0 = 51.3\ amu$$

$$^{57}_{26}\text{Fe} \quad 2.17\% \quad 56.9\,amu$$

$$^{57}_{26}\text{Fe} \ .0217 \times 56.9 = 1.23\ amu$$

$$^{58}_{26}\text{Fe} \quad 0.31\% \quad 57.9\,amu$$

$$^{58}_{26}\text{Fe} \ .0031 \times 57.9 = \underline{.179\ amu}$$

$$55.859$$
$$55.9\ amu$$

Calculate the atomic weight of iron.   (Work problem in the space provided).

40.  Calculate the formula weight of each of the following compounds.

d.  chromium (III) sulfate, $Cr_2(SO_4)_3$

$$FW = n_1W_1 + n_2W_2 + n_3W_3 + \ldots$$

where $n_1$ is the number of atoms of one element in the formula unit and $W_1$ is its atomic weight.

It is helpful at the beginning to consider formulas with parentheses in the following way:

$$Cr_2(SO_4)_3 = Cr, Cr, SO_4, SO_4, SO_4$$

$$
\begin{aligned}
FW \text{ of } Cr_2(SO_4)_3 \;=\; & (2 \text{ atoms Cr)(atomic weight of Cr)} \\
& +(3 \text{ atoms S) (atomic weight of S)} \\
& +(12 \text{ atoms O)(atomic weight of O)} \\[4pt]
=\; & (2 \text{ atoms Cr}) \left( \frac{52.0\,amu}{1 \text{ atom Cr}} \right) + (3 \text{ atoms S}) \left( \frac{32.1\,amu}{1 \text{ atom S}} \right) \\[4pt]
& +(12 \text{ atoms O}) \left( \frac{16.0\,amu}{1 \text{ atom O}} \right) \\[4pt]
=\; & 104.0\,amu + 96.3\,amu + 192.0\,amu \\[4pt]
=\; & 392.3\,amu
\end{aligned}
$$

Consult Appendix C in the text for answers to parts a, b, and c.

44.  Table sugar is sucrose, $C_{12}H_{22}O_{11}$.   Calculate the percentage composition of sucrose.

First find the formula weight of sucrose.

$$\text{FW } C_{12}H_{22}O_{11} = (12 \text{ atoms } C)\left(\frac{12.0\,\text{amu}}{1 \text{ atom } C}\right) + (22 \text{ atoms } H)\left(\frac{1.0\,\text{amu}}{1 \text{ atom } H}\right)$$

$$+ (11 \text{ atoms } O)\left(\frac{16.0\,\text{amu}}{1 \text{ atom } O}\right)$$

$$= 144\,\text{amu} + 22\,\text{amu} + 176\,\text{amu}$$

$$= 342\,\text{amu}$$

Then to find the percentage of each element, divide the total mass of that element in the compound by the FW and multiply by 100%.  Notice that the mass of each element in the compound is shown in the calculation above.

$$\%C = \left(\frac{\text{mass of C in } C_{12}H_{22}O_{11}}{\text{FW of } C_{12}H_{22}O_{11}}\right)(100\%) = \left(\frac{144\,\text{amu}}{342\,\text{amu}}\right)(100\%) = 42.1\%$$

$$\%H = \left(\frac{\text{mass of H in } C_{12}H_{22}O_{11}}{\text{FW of } C_{12}H_{22}O_{11}}\right)(100\%) = \left(\frac{22\,\text{amu}}{342\,\text{amu}}\right)(100\%) = 6.4\%$$

$$\%O = \left(\frac{\text{mass of O in } C_{12}H_{22}O_{11}}{\text{FW of } C_{12}H_{22}O_{11}}\right)(100\%) = \left(\frac{176\,\text{amu}}{342\,\text{amu}}\right)(100\%) = 51.5\%$$

The sum of the percentages adds up to 100%.

$$42.1\% + 6.4\% + 51.5\% = 100.0\%$$

***Practice Problem D

Calculate the percentage composition of ammonium phosphate, $(NH_4)_3PO_4$.

<hr>



Circle the correct answer in the following multiple choice questions.

1.  The complete symbol for an atom with 50 protons and 70 neutrons is

    a.   $^{70}_{50}Sn$

    b.   $^{50}_{70}Sn$

    c.   $^{120}_{50}Sn$

    d.   $^{120}_{70}Yb$

2.  How many protons, neutrons and electrons are found in $^{15}_{7}N^{3-}$?

    a.   7 protons, 8 neutrons, 7 electrons
    b.   7 protons, 8 neutrons, 3 electrons
    c.   7 protons, 8 neutrons, 4 electrons
    d.   7 protons, 8 neutrons, 10 electrons

3.  Which of the following statements is not true for $^{1}_{1}H$?

    a.   The atomic number 1, tells the number of protons.
    b.   $^{1}_{1}H$ is a neutral atom.
    c.   There is 1 proton, 1 neutron and 1 electron in $^{1}_{1}H$.
    d.   There are no neutrons in $^{1}_{1}H$.

4.  Which of the following pairs are isotopes of the same element?

    a.   $^{4}_{2}X$,   $^{2}_{4}X$

    b.   $^{4}_{2}X$,   $^{3}_{2}X$

    c.   $^{4}_{2}X$,   $^{4}_{3}X$

    d.   $^{4}_{2}X$,   $^{8}_{4}X$

5.  Which of the following pairs is isoelectronic?

    a.   $Al^{3+}$, $Na^{+}$
    b.   $Cu^{+}$, $Cu^{2+}$
    c.   Kr, Ar
    d.   $Fe^{2+}$, $Cu^{2+}$

6.  Copper has two naturally occurring isotopes, 69.09% is $^{63}_{29}Cu$ with a mass of 62.9amu and 30.91% is $^{65}_{29}Cu$ with a mass of 64.9amu.  Which of the following solutions is correct for calculating the atomic weight of copper?

    a.   (69.09 X 62.9amu) + (30.91 X 64.9amu) =
    b.   (62.9amu + 64.9amu) ÷ 2 =
    c.   (.6909 X 63amu) + (.3091 X 65amu) =
    d.   (.6909 X 62.9amu) + (.3091 X 64.9amu) =

7.  Which of the following formulas has the correct number of nitrogen atoms written beside it.

    a.   $(NH_4)_3PO_4$, 12 nitrogen atoms

    b.   $Ca(NO_3)_2$, 2 nitrogen atoms

    c.   $Be(CN)_2$, 1 nitrogen atom

    d.   $NH_4NO_3$, 1 nitrogen atom

8.  Which of the following compounds has a formula weight of 85.0g?
    a.   $NaNO_3$

    b.   $CaCl_2$

    c.   HCN

    d.   $Al_2(SO_4)_3$

9.  For the compound $CaCO_3$, which of the following solutions is correct for the calculation of percentage oxygen?

    a.   $\left( \dfrac{\text{mass of O in } CaCO_3}{\text{FW of } CaCO_3} \right) =$

    b.   $\left( \dfrac{48.0amu}{100.1amu} \right) =$

    c.   $\left( \dfrac{48.0amu}{100.1amu} \right) (100\%) =$

    d.   $\left( \dfrac{16.0amu}{100.1amu} \right) (100\%) =$

10.  All of the following statements about atoms are true except one.  Identify the incorrect statement.

    a.   Atoms are electrically neutral.
    b.   Atoms form cations by gaining protons.
    c.   All atoms of the same element have the same number of protons.
    d.   The mass of an atom is concentrated in the nucleus.

## ANSWERS TO PRACTICE PROBLEMS

Practice Problem A

| Symbol | Atomic Number | Mass Number | Protons | Neutrons | Electrons |
|---|---|---|---|---|---|
| $^{25}_{13}Al$ | 13 | 25 | 13 | 12 | 13 |
| $^{32}_{17}Cl$ | 17 | 32 | 17 | 15 | 17 |
| $^{39}_{19}K^{+}$ | 19 | 39 | 19 | 20 | 18 |
| $^{34}_{16}S^{2-}$ | 16 | 34 | 16 | 18 | 18 |

Practice Problem B

    a.  $Cl^{-}$, Cl gains $1e^{-}$ to achieve 18

    b.  $Ca^{2+}$, Ca loses $2e^{-}$ to achieve 18

    c.  $P^{3-}$, P gains $3e^{-}$ to achieve 18

    d.  $K^{+}$, K loses $1e^{-}$ to achieve 18

Practice Problem C

$^{54}_{26}Fe$     $.0584 \times 53.9\,amu = 3.15\,amu$

$^{56}_{26}Fe$     $.9168 \times 56.0\,amu = 51.3\,amu$

$^{57}_{26}Fe$     $.0217 \times 56.9\,amu = 1.23\,amu$

$^{58}_{26}Fe$     $.0031 \times 57.9\,amu = \underline{\quad.179\,amu}$

                                              $= 55.859\,amu$

                                              $= 55.9\,amu$

Notice that the average mass is closest to the mass of the most abundant isotope.

Practice Problem D

$$FW\ (NH_4)_3PO_4 = (3\ atoms\ N)\left(\frac{14.0amu}{1\ atom\ N}\right)$$

$$+(12\ atoms\ H)\left(\frac{1.0amu}{1\ atom\ H}\right)$$

$$+(1\ atom\ P)\left(\frac{31.0amu}{1\ atom\ P}\right)$$

$$+(4\ atoms\ O)\left(\frac{16.0amu}{1\ atom\ O}\right)$$

$$= 42.0amu + 12\ amu + 31.0amu + 64.0amu$$

$$= 149amu$$

$$\%N = \left(\frac{42.0amu}{149.0amu}\right)(100\%) = 28.2\%\ N$$

$$\%H = \left(\frac{12amu}{149.0amu}\right)(100\%) = 8.0\%\ H$$

$$\%P = \left(\frac{31.0amu}{149.0amu}\right)(100\%) = 20.8\%$$

$$\%O = \left(\frac{64.0amu}{149.0amu}\right)(100\%) = 43.0\%$$

The sum of the percentages adds up to 100%.

$$28.2\% + 8.0\% + 20.8\% + 43.0\% = 100.0\%$$

---

## ANSWERS TO SELF-TEST

1.  c
2.  d
3.  c
4.  b
5.  a
6.  d
7.  b
8.  a
9.  c
10.  b

# ELECTRON ARRANGEMENTS IN ATOMS

## CHAPTER
# 4

---

SUMMARY OUTLINE

---

### 4.1 Early Models of the Atom

The Thomsen plum pudding model, proposed in 1902, described the atom as a sphere containing uniformly distributed positive charges with electrons stuck in (Figure 4-1).

The Rutherford model, proposed in 1911, suggested that electrons move in orbits about the nucleus (Figure 4-2).

### 4.2 The Nature of Light

Radiant energy moves in waves.

Wavelength, $\lambda$, is the distance between two equivalent points in the path traveled by the wave (Figure 4-4).

Light is radiant energy having wavelengths in the range $10^{-8}$ to $10^{-6}$ m (Figure 4-5).

Visible light has wavelengths between 380nm and 780nm (1nm = $10^{-9}$m).

Each specific color of visible light has its own specific wavelength.

The energy of light is given by the equation

$$E = \frac{hc}{\lambda}$$

where h and c are constants.  As the wavelength of light
increases, the energy decreases.

### 4.3   Emission Spectra of the Elements

When light of all wavelengths passes through a prism, a continuous spectrum is produced.

Light emitted by a hot, gaseous element produces an emission
spectrum in which only certain wavelengths appear.  These
atomic emission spectra are called line spectra.  Each line
represents a specific wavelength and energy.

### 4.4   The Bohr Atom

Rutherford's model incorrectly predicted continuous spectra
for atoms.

The Bohr model, in accounting for the observed line spectra,
proposed that electrons exist in specific energy states in
atoms.  Transitions between energy states involve a definite
amount, or quantum, of energy.

### 4.5   Modern Ideas of Atomic Structure

DeBroglie proposed that matter has wavelike properties.

The Hersenberg uncertainty principle states that it is not
possible to know both the position and energy of an electron.
Since the specific energy state of the electron is known, the
location is uncertain.

An atomic orbital is a volume of space in which the electron
is likely to be found (Figure 4-10).

### 4.6   Energy Levels of Electrons

Electrons in a atom are distributed among shells.  Each shell
is a principal energy level designated by the symbol n.

The higher the value of n, a positive whole number, the
higher the energy of the electrons in the shell and the
larger the volume of the shell.

Each shell can hold $2n^2$ electrons (Table 4-1).

### 4.7   Energy Sublevels

Each shell consists of one or more energy sublevels.  The
number of sublevels in an energy level is the same as the
value of n.

Each sublevel consists of one or more atomic orbitals, designated as s, p, d, f, g, h, ... orbitals.  The s atomic
orbital is lowest in energy, followed by p orbitals, then d
orbitals, and so forth.

Each type of atomic orbital has its own characteristic shape (Figures 4-11, 12, 13).

Each atomic orbital can hold no more than two electrons.

## 4.8  Electron Configurations

The arrangement of electrons in an atom is called the electron configuration of the atom.

Electrons occupy the shells and atomic orbitals of lowest energy.

The notation for the electron configuration of lithium is

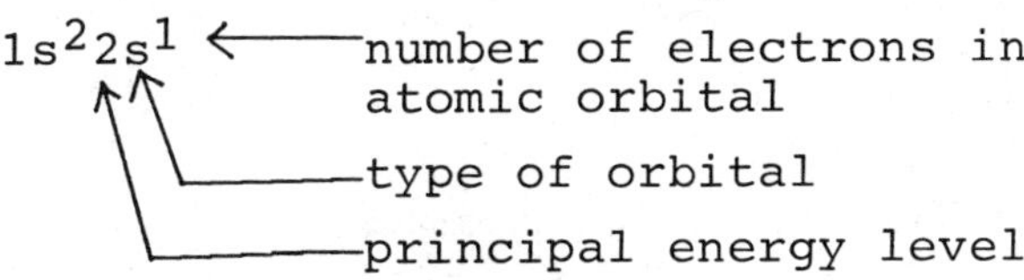


A diagram can be drawn to develop the electron configuration for any atom (Figure 4-17).

## 4.9  Quantum Numbers (Optional)

A mathematical description of the atom, called quantum mechanics, uses four quantum numbers to locate each electron in the atom.

The principal quantum number, n, gives the energy level.

$$n = 1, 2, 3, \ldots$$

The orientation quantum number, $\ell$, gives the type of orbital

$$\ell = 0, 1, 2, \ldots, n-1$$

The number of values of the magnetic quantum number, $m_\ell$, indicates the number of orbitals (or orientations) of each orbital type.

$$m_\ell = 0, \pm 1, \pm 2, \ldots, \pm \ell$$

The spin quantum number, $m_s$, refers to the 2 possible spins of the electron.

$$m_s = +\frac{1}{2} \text{ or } -\frac{1}{2}$$

Electron configurations can be described by use of quantum numbers (Table 4-5).

---

## SOLUTIONS TO SELECTED STUDY QUESTIONS AND PROBLEMS
### and
### PRACTICE PROBLEMS

7.  Describe how the energy of light is related to its wavelength. Does infrared light have more energy than ultraviolet light? Explain your answer.

The equation $E = \dfrac{hc}{\lambda}$ relates the energy of light to its wavelength.  As $\lambda$ increases, the value of $\dfrac{hc}{\lambda}$ decreases, so E decreases.  Thus energy and wavelength are inversely related.

In Figure 4-5, it can be seen that infrared radiation has a longer wavelength ($\sim 10^{-5}$m) than ultraviolet radiation ($\sim 10^{-8}$m). Thus, infrared radiation must have lower energy than ultraviolet radiation.

25.   How does one calculate the electron capacity of a shell? What is the electron capacity of each of the following shells?

    a.  1        b.  2        c.  3        d.  4

The electron capacity of a shell, or energy level n, is given by $2n^2$.

a.    $2n^2 = 2(1)^2 = \phantom{0}2$ electrons

b.    $2n^2 = 2(2)^2 = \phantom{0}8$ electrons

c.    $2n^2 = 2(3)^2 = 18$ electrons

d.    $2n^2 = 2(4)^2 = 32$ electrons

27.   How many energy sublevels are in a particular principal energy level?

Energy level n has n sublevels.

Energy level 1, n = 1, has 1 sublevel (s).
Energy level 2, n = 2, has 2 sublevels (s and p).
Energy level 3, n = 3, has 3 sublevels (s, p and d).
Energy level 4, n = 4, has 4 sublevels (s, p, d and f).
Energy level 5, n = 5, has 5 sublevels (s, p, d, f and g).

Higher energy levels have additional sublevels whose letters continue in alphabetical order from g.

***Practice Problem A

Identify the atomic orbitals in the fifth energy level.  Which orbital is highest in energy?  How many total atomic orbitals and electrons will the fifth energy level accommodate? (Answer problem in the space provided).

35.   What is the maximum number of s electrons that can exist in
      any one principal energy level?  How many p electrons?  How
      many d electrons?  How many f electrons?  Explain your
      answer.

      First remember that an atomic orbital can accommodate a
      maximum of 2 electrons.  Then recall the number of orbitals
      possible for each orbital type.  The answer is given in the
      following table.

| Type of sublevel | Number of orbitals possible | Maximum number of electrons |
|---|---|---|
| s | 1 | 2 |
| p | 3 | 6 |
| d | 5 | 10 |
| f | 7 | 14 |

***Practice Problem B

   How many electrons are present in the third energy level?
   Identify the sublevel and number of electrons in each sublevel.  (Answer problem in the space provided).

37.   Using only the periodic table or list of elements, write the
      electron configuration of each of the following atoms.

      d.  V

         From the periodic table, vanadium's atomic number of 23
         indicates 23 electrons in the atom.

         Draw a diagram like Figure 4-17.  Be sure that you can do
         this without reference to the text, since the text may not
         be available during testing.  In addition to the diagram
         remember the total electron capacities of each orbital
         type:

                s (1 orbital),   2 electrons
                p (3 orbitals),  6 electrons
                d (5 orbitals), 10 electrons
                f (7 orbitals), 14 electrons

      Until you are thoroughly familiar with the orbitals and
      their electron capacities, it may be helpful to write the

above listing along with the diagram on your paper.  It's
not necessary to memorize the listing if you notice the
odd number relationship for s, p, d, f orbitals and
remember that each orbital can hold a maximum of 2e⁻.
For vanadium, the 23e⁻ are arranged in the atom in the
following way:

$$1s^2 2s^2 2p^6 3s^2 3p^6 4s^2 3d^3 \text{ or } [Ar] 4s^2 3d^3$$

Adding up the superscripts in the complete electron configuration results in 23, the number of e⁻ in V.  The
short form, $[Ar] 4s^2 3d^3$, is equivalent to the longer configuration.  Ar, with 18e⁻, is the element at the end of
the previous period.  It has the configuration
$1s^2 2s^2 2p^6 3s^2 3p^6$.  [Ar] can be written in place of this
configuration.  The electron configuration of the 5
additional electrons found in vanadium are written to the
right of [Ar].

f.   Re

Rhenium has 75 electrons.  Although the electron configuration for Re is much longer than vanadium's, it is
not more difficult to write.  Use the orbital filling
diagram and electron capacity list to arrange the 75
electrons.

$$1s^2 2s^2 2p^6 3s^2 3p^6 4s^2 3d^{10} 4p^6 5s^2 4d^{10} 5p^6 6s^2 4f^{14} 5d^5 \text{ or}$$

$$[Xe] 6s^2 4f^{14} 5d^5$$

To check, add the superscripts in the complete e⁻ configuration.  The total is 75.

38.   Using only the periodic table or list of elements, write the
electron configuration of each of the following ions.

a.   I⁻

I has 53e⁻, I⁻ has 54e⁻.

Again, draw a diagram for the order of filling orbitals
and write the electron capacities of each orbital type.
From this, assign the 54e⁻ in I⁻.

$$1s^2 2s^2 2p^6 3s^2 3p^6 4s^2 3d^{10} 4p^6 5s^2 4d^{10} 5p^6$$

To write the short form, the text states that the atomic
symbol of the element at the end of the immediately
preceding period is shown in brackets, followed by
symbols for orbitals present beyond those of the elements
in brackets.  This is true for neutral atoms.  If the
short form of the electron configuration for I⁻ is written
this way, it becomes

$$[Kr] 5s^2 4d^{10} 5p^6 .$$

Notice however that Xe has 54e⁻ and $[Kr] 5s^2 4d^{10} 5p^6$ can be
further shortened to [Xe].  For negative ions, the

shortest form of the electron configuration is accomplished by writing the symbol of the element at the end of the same period.

b.  $Ra^{2+}$

Ra has $88e^-$, $Ra^{2+}$ has $86e^-$.

The electron configuration is
$1s^2 2s^2 2p^6 3s^2 3p^6 4s^2 3d^{10} 4p^6 5s^2 4d^{10} 5p^6 6s^2 4f^{14} 5d^{10} 6p^6$

The short form is [Rn] which also has $86e^-$.  Notice that $Ra^{2+}$ and Rn have the same number of electrons and the same electron configurations.

Consult Appendix C in the text for answers to parts c and d.

***Practice Problem C

Write the electron configurations and short form electron configurations for the following elements or ions.  (Answer questions in the space provided).

a.  As   $[Ar] 4s^2 3d^{10} 4p^3$

b.  Tc   $[Kr] 5s^2 4d^5$

c.  $P^{3-}$   $[Ar]$

d.  $Cs^+$   $[Xe]$

42.  What are the values of $m_\ell$ for a d orbital?  How many orbitals are there?

The values of $m_\ell$ are
$$m_\ell = +\ell, +(\ell-1), +(\ell-2), \ldots, 0, \ldots -(\ell-2), -(\ell-1), -\ell$$

where $\ell$, is the orbital type quantum number.  For the d orbital, the value of $\ell$ is 2.

For $\ell = 2$,

$$m_\ell = +\ell = 2$$

$$m_\ell = +(\ell-1) = +(2-1) = 1$$

$$m_\ell = +(\ell-2) = +(2-2) = 0$$

$$m_\ell = -(\ell-1) = -(2-1) = -1$$

$$m_\ell = -\ell = -2$$

or $m_\ell = -2, -1, 0, +1, +2$

Another way to express the allowed values for $m_\ell$ is integers that range from the negative value of $\ell$ through zero up to the positive value of $\ell$.

$$m_\ell = 0, \pm 1, \pm 2, \ldots, \pm \ell$$

or, in this case,

$$m_\ell = 0, \pm 1, \pm 2$$

Since there are 5 values for $m_\ell$, there are 5 spatial orientations, or simply 5 d orbitals.

44.    How many orbitals are there in the n = 4 energy level?

The types of orbitals are determined by the quantum number $\ell$.

$$\ell = 0, 1, 2, 3, \ldots, n-1$$

For n = 4,

$$\ell = 0, 1, 2, 3, \longrightarrow s, p, d, f \text{ orbitals}$$

(The maximum value of $\ell$ is n-1 = 4-1 = 3).

The number of orbitals of each type is determined by the quantum number $m_\ell$.

$$m_\ell = 0, \pm 1, \pm 2, \ldots, \pm \ell$$

There are 4 values of $\ell$ or 4 types of orbitals.  The number of orbitals of each type must be found.

For $\ell = 0$ (s orbital)

$$m_\ell = 0, \text{ 1 value indicates 1 orbital}$$

For $\ell = 1$ (p orbital)

$$m_\ell = 0, \pm 1 \text{ or } -1, 0, +1$$

3 values indicate 3 orbitals

For $\ell = 2$ (d orbital)

$$m_\ell = 0, \pm 1, \pm 2 \text{ or } -2, -1, 0, +1, +2$$

5 values indicate 5 orbitals

For  $\ell = 3$ (f orbital)

$m_\ell$ = 0, ±1, ±2, ±3 or -3, -2, -1, 0, +1, +2, +3

7 values indicate 7 orbitals.

Add up the number of orbitals of each type.

1 + 3 + 5 + 7 = 16 orbitals

There are 16 orbitals in the n = 4 energy level.

***Practice Problem D

What are the allowed quantum numbers in the following situations?  What do the quantum numbers indicate about the atom represented?  (Answer questions in the space provided).

a.   values of $\ell$ when n = 3

$\ell = 0, 1, 2$

b.   values of $m_\ell$ when $\ell = 1$

$M_L = 0 \pm 1$

c.   values of $m_\ell$ for s orbitals

$M_\ell = 0$

45.   Each of the following sets of quantum numbers (n, $\ell$, $m_\ell$, $m_s$) describes an electron in an atom.  Give the principal energy level and orbital designated by each.

a.   3, 2, 1, $-\dfrac{1}{2}$

Since n = 3 and $\ell = 2$, the electron is located in a 3d orbital.  The $m_\ell$ and $m_s$ quantum numbers are necessary to distinguish between all 10e$^-$ that could occupy d orbitals in the 3rd energy level.  $m_\ell$ values are not necessary to identify the energy level and type of orbital.

c.   5, 1, -1, $+\frac{1}{2}$

Since n = 5, and $\ell$ = 1, the electron is located in a 5p orbital.

Consult Appendix C in the text for answers to parts b and d.

46.   Using quantum numbers, explain the absence of d orbitals in the second principal energy level.

For d orbitals, $\ell$ = 2.  The allowed values of $\ell$ in energy level n is

$$\ell = 0, 1, 2, \ldots n-1$$

In the second energy level (n = 2), $\ell$ = 0, 1.

One is the maximum value for $\ell$ since n-1 = 2-1 = 1.  Therefore, only s orbitals ($\ell$ = 0) and p orbitals ($\ell$ = 1) are allowed in the second energy level.

***Practice Problem E

Write the complete set of quantum numbers for all electrons that could be accommodated by the 2p subshell.  (Answer the question in the space provided).

---

## SELF-TEST

Circle the correct answer in the following multiple choice questions.

1.   The Rutherford model of the atom suggested that

    a.   the nucleus contains most of the atom's mass and volume.
    b.   the electrons occupy little volume in the atom.
    c.   the nucleus has no charge but most of the atom's mass.
    d.   the nucleus contains all of the positive charge, most of the mass and very little of the volume of the atom.

2.  Which of the following wavelengths has the highest energy?

    a.  380nm
    b.  $4 \times 10^{-11}$m
    c.  600nm
    d.  $5.7 \times 10^{-3}$m

3.  The atomic emission spectrum of an element is

    a.  a continuous spectrum.
    b.  characteristic of all elements in one group on the
        periodic table.
    c.  a line spectra.
    d.  more complex for elements with low atomic numbers.

4.  Which of the following statements is incorrect for the Bohr
    model of the atom?

    a.  Atoms in their ground states are unstable, since energy
        is continually lost as electrons orbit the nucleus.
    b.  Electrons in low energy states are closer to the nucleus
        than electrons in high energy states.
    c.  Only certain well-defined energies are allowed for
        electrons.
    d.  When an atom absorbs energy, electrons go from the
        ground state to an excited state.

5.  Which of the following is not true for the modern atomic
    model?

    a.  Electrons behave as both particles and waves.
    b.  The energy and location of an electron is known.
    c.  The region of space in which an electron can be found is
        called an atomic orbital.
    d.  An electron can have only certain specific energies.

6.  Energy level 3

    a.  is lower in energy than energy level 2.
    b.  is farther from the nucleus than energy level 4.
    c.  can accommodate 18 electrons.
    d.  has 2 sublevels.

7.  Which of the following orientation quantum numbers correctly
    identifies the orbital type?

    a.  $\ell = 1$, s orbital
    b.  $\ell = 2$, f orbital
    c.  $\ell = 2$, p orbital
    d.  $\ell = 4$, g orbital

8.  What is the correct electron configuration for Ga?
    a.  $1s^2 2s^2 2p^6 3s^2 3p^6 4s^2 4p^6 5s^1$
    b.  $1s^2 2s^2 2p^6 3s^2 3p^6 4s^2 3d^{10} 4p^1$
    c.  $[Ar] 4s^2 4p^1$
    d.  $1s^2 2s^2 2p^6 3s^2 3p^1$

54   Introduction to Chemistry

9.    If n = 3, then

   a.   $\ell = 2,\ m_\ell = 3$

   b.   $\ell = 1,\ m_\ell = 2$

   c.   $\ell = 2,\ m_\ell = -2$

   d.   $\ell = 3,\ m_\ell = -3$

10.   A set of possible quantum numbers $(n,\ \ell,\ m_\ell,\ m_s)$ for an
      electron in a 4f orbital is

   a.   $4,\ 3,\ 2,\ +\dfrac{1}{2}$

   b.   $4,\ 4,\ 0,\ +\dfrac{1}{2}$

   c.   $3,\ 3,\ -2,\ -\dfrac{1}{2}$

   d.   $4,\ 3,\ -4,\ +\dfrac{1}{2}$

---

## ANSWERS TO PRACTICE PROBLEMS

**Practice Problem A**

   when n = 5, there are 5 sublevels, s, p, d, f and g atomic
   orbitals.  The g atomic orbital is highest in energy.

   For sublevel s, there is 1 atomic orbital.
   For sublevel p, there are 3 atomic orbitals.
   For sublevel d, there are 5 atomic orbitals.
   For sublevel f, there are 7 atomic orbitals.
   For sublevel g, there are 9 atomic orbitals.

   The total number of orbitals is 25.  Since each orbital can
   hold no more than 2 electrons, 50 electrons can be accommodated.
   This follows from the $2n^2$ rule; when n = 5, $2n^2 = 2(5)^2 = 50$
   electrons.

**Practice Problem B**

   For n = 3, $2n^2 = 2(3)^2 = 18$ electrons

   The third energy level has 3 sublevels, s, p, and d.

   | s, 1 orbital,  | 2 electrons  |
   |----------------|--------------|
   | p, 3 orbitals, | 6 electrons  |
   | d, 5 orbitals, | 10 electrons |

**Practice Problem C**

   a.   $1s^2 2s^2 2p^6 3s^2 3p^6 4s^2 3d^{10} 4p^3$ or $[Ar]4s^2 3d^{10} 4p^3$

   b.   $1s^2 2s^2 2p^6 3s^2 3p^6 4s^2 3d^{10} 4p^6 5s^2 4d^5$ or $[Kr]5s^2 4d^5$

   c.   $1s^2 2s^2 2p^6 3s^2 3p^6$ or $[Ar]$

   d.   $1s^2 2s^2 2p^6 3s^2 3p^6 4s^2 3d^{10} 4p^6 5s^2 4d^{10} 5p^6$ or $[Xe]$

Practice Problem D

   a.   When $n = 3$, $\ell = 0, 1, 2$

        These quantum numbers specify 3s, 3p, and 3d orbitals in
        an atom.

   b.   When $\ell = 1$, $m_\ell = 0, \pm 1$ or $-1, 0, +1$

        Since there are 3 values for $m_\ell$, these quantum numbers
        specify 3 orientations for p orbitals.

   c.   For an s orbital, $\ell = 0$

        The only value allowed for $m_\ell$ is 0.  Since there is only
        one value for $m_\ell$, there is only one orientation for an
        s orbital.

Practice Problem E

   The 2 in 2p indicates $n = 2$, the 2nd energy level; the p
   orbital indicates $\ell = 1$.  When $\ell = 1$, $m_\ell = -1, 0, +1$

   The complete set of quantum numbers ($n$, $\ell$, $m_\ell$, $m_s$) for all
   electrons in the 2p subshell is

$$\left. \begin{array}{l} 2,\ 1,\ -1,\ +\dfrac{1}{2} \\[1em] 2,\ 1,\ -1,\ -\dfrac{1}{2} \end{array} \right\} \text{one orbital}$$

$$\left. \begin{array}{l} 2,\ 1,\ 0,\ +\dfrac{1}{2} \\[1em] 2,\ 1,\ 0,\ -\dfrac{1}{2} \end{array} \right\} \text{one orbital}$$

$$\left. \begin{array}{l} 2,\ 1,\ 1,\ +\dfrac{1}{2} \\[1em] 2,\ 1,\ 1,\ -\dfrac{1}{2} \end{array} \right\} \text{one orbital}$$

   The three $m_\ell$ values specify three p orbitals.  Since each
   orbital can hold $2e^-$, a total of $6e^-$ are allowed.  Six sets
   of quantum numbers are given to show six $e^-$.

   Notice that 2 sets of quantum numbers can have identical
   $n$, $\ell$, $m_\ell$ quantum numbers.  These refer to the same p orbital.
   The two different values for $m_s$, $+\dfrac{1}{2}$ and $-\dfrac{1}{2}$, show 2
   different $e^-$ in that orbital.

---

ANSWERS TO SELF-TEST

   1.   d
   2.   b
   3.   c
   4.   a
   5.   b
   6.   c
   7.   d
   8.   b
   9.   c
  10.   a

---

## SUMMARY OUTLINE

**5.1   The Periodic Law**

The periodic law, proposed by Mendeleev in 1869, states that elements arranged in order of increasing atomic weights will show a periodic change in properties.

It is now known that chemical periodicity is a function of atomic number rather than atomic weight.

**5.2   The Modern Periodic Table**

The seven horizontal rows on the periodic table are called periods.

Vertical columns are called groups.

Each period, beginning with an element in group IA, marks the filling of electrons into a new principal energy level.  Each consecutive element in a period has one more electron in its atom.

All elements in a group have identical outer energy level electron configurations, except for the principal energy level.  Since outer shell electron configurations account for an element's chemical properties, all elements in a group will have similar chemical properties.

The outer principal energy level is called the valence shell. Electrons in the valence shell are called valence electrons.

The periodic table is divided by a zigzag line into metals, on the left, and nonmetals, on the right.  Elements bordering each side of the zigzag line are called metalloids.

Metal atoms tend to have 1-3 valence electrons.  Nonmetal atoms tend to have 4-8 valence electrons.

Atoms of Group IA and IIA elements have only S electrons in their valence shell.  Electrons are being filled into p orbitals in Groups IIIA-VIIA and Group 0.  Groups IB-VIIIB have atoms filling electrons in d orbitals.  f orbitals are being filled in elements 58-71 and 90-103 (Figure 5-3).

## 5.3   Periods of Elements

The number of elements in each period is determined by the electron capacities of the orbitals being filled (Table 5-1).

## 5.4   Groups of Elements

Group A and Group 0 elements, called representative elements, have electrons filling in s and p orbitals (Figure 5-4).

For all representative elements except Group 0, the group number gives the number of valence electrons.

Group 0 elements, except He, have 8 valence electrons.

Elements in B groups, called transition elements, are filling inner orbitals.

## 5.5   A Survey of Representative Elements

Group IA elements are called alkali metals.  Their valence shell electron configuration is $ns^1$.

Hydrogen has an electron configuration of $1s^1$ and is a Group IA element, but it is not an alkali metal. Hydrogen is a nonmental.

Group IIA elements, called alkaline earth metals, have a valence shell electron configuration of $ns^2$.

Group IIIA elements have a valence shell electron configuration of $ns^2np^1$.

Group IVA elements have a valence shell electron configuration of $ns^2np^2$.

Group VA elements have a valence shell electron configuration of $ns^2np^3$.

Group VIA elements have a valence shell electron configuration of $ns^2np^4$.

Group VIIA elements, known as the halogens, have a valence shell electron configuration of $ns^2np^5$.

Group 0 elements, known as noble gases, have a valence shell electron configuration of $ns^2np^6$. Helium is an exception with only 2 electrons and a valence shell electron configuration of $1s^2$.

An atom or ion is least reactive when it contains eight valence electrons, or an octet of valence electrons. This octet rule applies to many elements and compounds.

To form cations, Group A metals lose all their valence electrons.

## 5.6  The Transition Elements

Transition elements include the elements in B groups and the lanthanides and actinides.

Transition metals form cations with at least two different positive charges. Electrons can be lost from the valence shell and from inner d and f orbitals.

## 5.7  Atomic Size

Atomic radii decrease across a period because the principal energy level remains the same while the number of protons in the nucleus increases (Figure 5-5).

Atomic radii increase going down a group due to the addition of new principal energy levels.

## 5.8  Ionization Energy

Ionization energy, (I.E.), is the energy required to remove the most loosely held electron from a gaseous ground-state atom or ion (Table 5-2).

$$A_{(g)} + energy \longrightarrow cation^+ + e^-$$

The energy to remove one electron is called the 1st I.E.; to remove a second electron, the 2nd I.E.; and so on.

First ionization energies increase across a period. Atoms become smaller enabling the nucleus to pull more tightly on the electrons (Figures 5-6 and 5-7).

First ionization energies decrease down a group. Nuclear attraction for valence electrons decreases as the atomic size increases.

As more electrons are removed from an atom or ion, the nuclear charge increases and the ionization energy increases.

It is very difficult to remove an electron from a filled octet.

Metals lose all of their valence electrons to achieve an octet in the shell just beneath the valence shell.

### 5.9  Electron Affinity

The energy change that occurs when an electron is added to a gaseous ground-state atom is called electron affinity.

$$A_{(g)} + e^- \longrightarrow anion^- + energy$$

Energy is usually released when an electron is added.

Electron affinities increase across a period and decrease down a group (Figure 5-8).  The nucleus of smaller atoms can attract electrons more strongly.

Nonmetals have a greater tendency to gain electrons than do metals.

Nonmetals gain electrons to achieve an octet in their valence shells.

### 5.10  The Chemical Behavior of Metals and Nonmetals

Metals have large sizes, low ionization energies and low electron affinities.

Nonmetals have small sizes, high ionization energies and high electron affinities.

Metals have a greater tendency than nonmetals to lose electrons.  Nonmetals have a greater tendency than metals to gain electrons (Figure 5-9).

Noble gases have small sizes; high ionization energies and low electron affinities.  The octet of electrons in noble gases account for their stability.

---

### SOLUTIONS TO SELECTED STUDY QUESTIONS AND PROBLEMS
### and
### PRACTICE PROBLEMS

---

11.  Using only the periodic table and Figure 5-3, give the valence shell electron configuration of each of the following atoms.

   a.  C

   b.  Ba

   c.  Ar

   d.  As

   e.  Cs

   f.  S

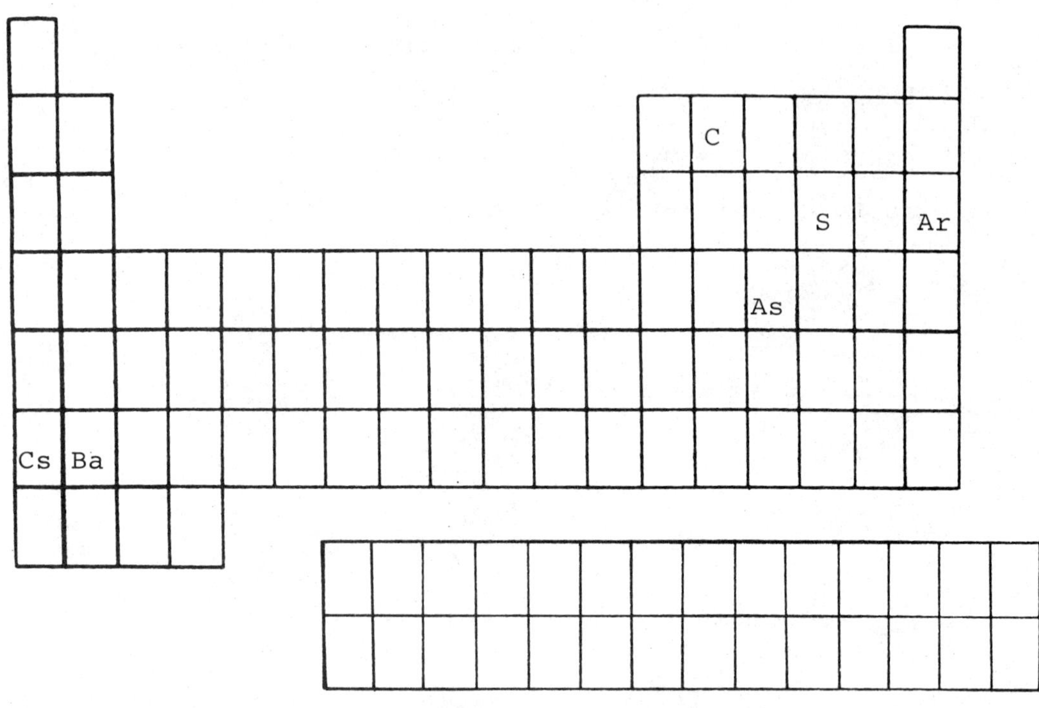


Figure 5-3

a.  Notice carbon's position on the periodic table in Figure
    5-3.  There are two things to consider:

    The horizontal row indicates the number of shells or
    principal energy levels occupied by electrons in the
    atom.  Carbon is in row 2, known as period 2, so the
    valence shell (outermost and highest in energy) is the
    second energy level.

    The vertical column indicates the type of orbital
    occupied by the highest energy electron.  Carbon is
    found in the second column of the p block.  This means
    that there are 2 electrons in p orbitals in the valence
    shell.

    Combining the above information determined from carbon's
    location results in an abbreviated electron configuration
    of $2p^2$.  This is incomplete since all electrons in a
    valence shell must be given.  Notice that period 2 also

has two columns in the s block that represents electrons
filling in the 2s orbital.  Including these electrons
gives the valence shell electron configuration for C as
$2s^2 2p^2$.

As a check, consult a complete periodic table.  Carbon is
in Group IVA which indicates 4 valence electrons having a
$ns^2 np^2$ configuration.

b.  Check Barium's location on the periodic table.

   The sixth row, known as period 6, indicates that the
   sixth energy level is the valence shell.

   The second column in the s block indicates 2 electrons
   in the s orbital.

   Barium's valence shell electron configuration is $6s^2$.  As
   a check, notice that barium is in Group IIA.  This
   indicates 2 valence electrons having an $ns^2$ configuration.

c.  Argon is found in row three and the sixth column of the p
   block.  This gives an abbreviated configuration of $3p^6$.
   Include all electrons in the valence shell.  Period three
   also contains 2 columns in the s block.  The valence shell
   electron configuration for argon is $3s^2 3p^6$.  As a check,
   argon is in Group 0 which has 8 valence electrons and a
   $ns^2 np^6$ configuration.

d.  Arsenic is found in period four and column three of the p
   block.  This gives an abbreviated configuration of $4p^3$.
   Additionally, period four contains 2 s block columns and
   10 d block columns.  This would result in a configuration
   of $4s^2 3d^{10} 4p^3$.  However, only the electrons in the outer
   energy level are considered valence electrons.  The $3d^{10}$
   electrons are not reported in the valence shell configuration for arsenic.  The valence shell configuration
   for As is $4s^2 4p^3$.

   As a check, notice As is found in Group VA, indicating 5
   valence electrons with a $ns^2 np^3$ configuration.

e.  Cesium, in period six and column one of the s block, has a
   valence shell electron configuration of $6s^1$.  Cesium's
   position in Group IA confirms one valence electron and an
   $ns^1$ configuration.

f.  Sulfur, in period three and column four of the p block,
   has a valence shell electron configuration of $3s^2 3p^4$.  As
   a Group VIA element, arsenic has valence electrons with a
   $ns^2 np^4$ configuration.

***Practice Problem A

Using only the periodic table and Figure 5-3, give the
following answers in the space provided:

a.  the valence shell electron configuration of In.

b.  the complete electron configuration for Mg.

c.  the short form electron configuration for Br.

15.  Hydrogen and helium are in period 1.  How are they alike?
     How are they different?

As period 1 elements, both have electrons in principal energy
level one and electrons in s orbitals.  They differ in their
number of electrons and electron configurations.  Hydrogen,
with 1 electron, has a $1s^1$ configuration.  Helium, with 2
electrons has a $1s^2$ configuration.  Because H and He are in
different groups due to their different valence electron
configurations, they differ in chemical properties as well.

***Practice Problem B

Carbon and silicon are in Group IVA.  How are they alike?
How are they different?  (Answer in the space provided).

26.  Give the valence shell configuration for each group of re-
     presentative elements.

Representative elements are those elements in Groups A and
Group 0.  For the A groups, the Roman Numeral indicates the
number of valence electrons.

Group IA,    1 valence e⁻, $ns^1$
Group IIA,   2 valence e⁻, $ns^2$
Group IIIA,  3 valence e⁻, $ns^2np^1$
Group IVA,   4 valence e⁻, $ns^2np^2$
Group VA,    5 valence e⁻, $ns^2np^3$
Group VIA,   6 valence e⁻, $ns^2np^4$
Group VIIA,  7 valence e⁻, $ns^2np^5$

Group 0 includes the noble gases which have 8 valence electrons and a $ns^2np^6$ configuration.  Helium, with only $2e^-$ is the only exception.  He has a valence shell configuration of $1s^2$.

37.    Which atom do you think has the smallest radius?  Which would you expect to have the largest radius?

A period one element should have the smallest radius since there is only one principal energy level.  The choice between H and He is somewhat difficult to make.  H, with only one $e^-$, is probably the smallest but trends indicate atomic size decreases from left to right across a period.  Trends would indicate that He is the smallest.  Since Figure 5-5 does not give data for Group 0 elements, a definitive answer can be found in a handbook.  H has a radius of 0.037nm and He has a radius of 0.050nm.  The presence of only $1e^-$ in H appears to be a more important factor than the stronger pull of 2 protons in He.  He also has electron-electron repulsions not found in H.  These repulsions could contribute to a slightly larger atom.

The largest atom should be found in period 7.  Since size decreases across a period, the largest element in period 7 is on the far left.  Francium is the largest element.

***Practice Problem C

Select the largest atom in the following pairs of atoms and explain your choice.  (Answer problems in the space provided).

a.  N and 0




b.  S and Se




c.  K and Cl

41.   In each of the following pairs, select the element that would
      be expected to have the higher first ionization energy.  Use
      only the periodic table.  Give reasons for your selection.

      a.  Cl or Al

          Cl would have the higher first ionization energy.  Cl,
          to the right of Al in period 3, is smaller than Al.  Since
          the outermost electron in Cl is nearer to the nucleus, it
          is held more tightly.  Thus, more energy would be re-
          quired to remove one electron in Cl.

      b.  As or Rb

          Since As is smaller than Rb, its outermost electron would
          be held more tightly and more energy would be required to
          remove it.  As would have a higher first ionization
          energy.

      Consult Appendix C in the text for answers to parts c and d.

49.   In each of the following pairs, select the element that would
      be expected to have the higher electron affinity.  Use only
      the periodic table.  Give reasons for your selections.

      a.  K or F

          Smaller atoms have a greater tendency to gain electrons
          because an added electron will be closer to the nucleus
          and attracted with a greater force.  Since F is smaller
          than K, it would have a higher electron affinity.

      b.  Ca or Se

          Se is smaller than Ca, attracts an electron to the nucleus
          with greater force, and has a higher electron affinity.

      Consult Appendix C in the text for answers to parts c and d.

59.   From your knowledge of trends in ionization energies, predict
      which of the following cations is likely to exist.  Explain
      your answer.

      a.  $Al^{4+}$

          Aluminum has an electron configuration of $1s^2 2s^2 2p^6 3s^2 3p^1$.
          Removal of 3 electrons results in a filled octet in the
          second principal energy level, $2s^2 2p^6$.  It would be very
          difficult to remove a fourth electron from this filled
          octet.  $Al^{4+}$ is very unlikely to exist.

      c.  $Rb^+$

          Rubidium has a valence shell configuration of $5s^1$.  To
          form $Rb^+$, one $e^-$ is lost, leaving a filled octet in the
          fourth energy level.  $Rb^+$ is likely to exist.

      Consult Appendix C in the text for answers to parts b, d, e
      and f.

***Practice Problem D

Predict the charge on the ion formed by each of the following elements.  Indicate the number of electrons lost or gained and explain.  (Answer problems in the space provided).

a.  Ca

b.  S

---

## SELF-TEST

1.  Which of the following statements is not true for the periodic table of elements.

    a.  The horizontal rows are called periods.
    b.  The vertical columns are called groups.
    c.  Elements in B groups are called representative elements.
    d.  Elements with atomic numbers 58-71 are called the lanthanides.

2.  Elements with similar chemical properties

    a.  are placed in the same period in the periodic table.
    b.  have the same number of valence electrons.
    c.  are oxygen, fluorine and hydrogen.
    d.  are called representative elements.

3.  Which of the following atoms has the correct valence shell configuration?

    a.  Si, $3s^2 3p^2$
    b.  Ba, $5s^2$
    c.  Te, $5s^2 5p^6$
    d.  Ga, $4s^2 3p^1$

4.     Electrons are being filled into d orbitals in

   a.   atoms of Group IA elements.
   b.   atoms of Group IIIA elements.
   c.   atoms of Group IB elements.
   d.   atoms of Group 0 elements.

5.     Which of the following ions is likely to exist?

   a.   $Te^{2+}$
   b.   $N^{2-}$
   c.   $Ca^{3+}$
   d.   $Cs^{+}$

6.     Which of the following statements is not true for transition
       elements?

   a.   Transition elements can form both positive and negative
        ions.
   b.   Transition elements can lose electrons from inner d and
        f orbitals when cations form.
   c.   Transition elements include all Group B elements plus
        the lanthanides and actinides.
   d.   Transition metals often form highly colored compounds.

7.     Which of the following atoms has the largest radius?

   a.   Mg
   b.   P
   c.   K
   d.   Na

8.     Of the following elements, the ionization energy is highest
       for

   a.   S
   b.   Cl
   c.   Br
   d.   K

9.     Which of the following statements is true for electron
       affinity?

   a.   Electron affinities increase down a group.
   b.   Electron affinity increases as atoms get smaller.
   c.   Metals have a higher electron affinity than nonmetals.
   d.   When an electron is added, energy is usually absorbed.

10.    Of the following elements, the most metallic one is

   a.   F
   b.   Ba
   c.   I
   d.   Cs

---

## ANSWERS TO PRACTICE PROBLEMS

**Practice Problem A**

a.  $5s^2 5p^1$

b.  $1s^2 2s^2 2p^6 3s^2$

The periodic table can be used to write complete electron configuration.  Since Mg is in period three, electrons from energy levels 1, 2, and 3 must be identified. Period one has only 2 s block columns ($1s^2$).  Period two has 2 s block columns and 6 p block columns ($2s^2 2p^6$). Magnesium is found in period three and column two of the s block ($3s^2$).

c.  $[Ar]4s^2 3d^{10} 4p^5$

The periodic table can be used to write the short form electron configuration.  Write the last element in the previous period in brackets ([Ar]).  Bromine is located in period four and column five of the p block ($4p^5$).  2 s block columns ($4s^2$) and 10 d block columns ($3d^{10}$) precede the p block in the fourth period.  Combine these configurations in the order encountered on the periodic table, i.e., $[Ar]4s^2 3d^{10} 4p^5$.

**Practice Problem B**

Carbon and silicon have four valence electrons and similar chemical properties.  Carbon has a valence shell configuration of $2s^2 2p^2$ and silicon has a valence shell configuration of $3s^2 3p^2$.  Silicon has one more principal energy level and is larger than carbon.

**Practice Problem C**

a.  N

Nitrogen and oxygen both have two principal energy levels, but nitrogen has seven protons while oxygen has eight.  The nucleus of oxygen attracts its electrons more strongly than the nitrogen nucleus does.  This results in a larger atom for nitrogen.

b.  Se

Selenium has four principal energy levels and sulfur has three.  Since the fourth energy level is farther from the nucleus than the third, selenium is a larger atom.

c.  K

Potassium has electrons in four principal energy levels. Chlorine has electrons in only three principal energy levels.  Potassium is larger than chlorine.

**Practice Problem D**

a.  $Ca^{2+}$  2 electrons lost

Metals tend to lose all of their valence electrons to achieve an octet in the shell just beneath the valence shell.

Calcium has an electron configuration of $1s^2 2s^2 2p^6 3s^2 3p^6 4s^2$. The $4s^2$ electrons are lost when the $Ca^{2+}$ cation forms, leaving an octet in the third energy level.

b. $S^{2-}$, 2 electrons gained

Nonmetals tend to gain electrons to achieve an octet in their valence shell.  Sulfur has an electron configuration of $1s^2 2s^2 2p^6 3s^2 3p^6$.  Two electrons are gained to fill the p orbitals and achieve an octet in the third energy level.

---

## ANSWERS TO SELF-TEST

1. c
2. b
3. a
4. c
5. d
6. a
7. c
8. b
9. b
10. d

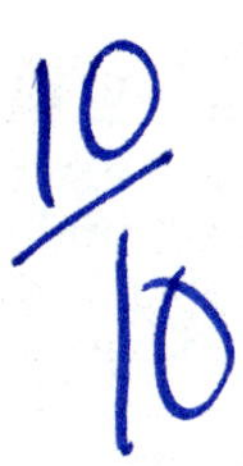

---

## SUMMARY OUTLINE

### 6.1   Types of Chemical Bonds

Compounds that conduct electricity in solution usually contain a metal and one or more nonmetals present as ions.  The force holding the ions together is called an ionic bond.  The compounds are called ionic compounds.

Compounds that do not conduct electricity in solution are composed of molecules.  These molecular compounds are usually composed of nonmetals only.  Forces that hold atoms together in a molecule are called covalent bonds.

### 6.2   Ionic Bonds

Metals lose their valence electrons to achieve a stable octet. Positive metal ions are called cations.

Nonmetals gain electrons to achieve a stable octet.  Negative nonmetal ions are called anions.

Cations and anions are usually isoelectronic with noble gases and more stable than their parent atoms.

Ionic compounds form when metal atoms lose electrons to nonmetal atoms.  The positively charged cations are attracted to the negatively charged anions.  The electrical attractions between oppositely charged ions are called ionic bonds (Figures 6-1 and 6-2).

A positive ion is smaller than its parent atom for two reasons:  (1) the valence energy shell is vacant, and (2) the remaining electrons are attracted more strongly to the nucleus.

A negative ion is larger than its parent atom.  The ion has the same number of principal energy levels as the atom, but the attraction from the nucleus is reduced by the addition of electrons.

## 6.3  Electron Dot Formulas of Atoms and Monatomic Ions

The electron dot formula of an atom represents the valence electrons as dots surrounding the element's symbol.  Unpaired electrons are represented by single dots, and paired electrons are represented by paired dots (Figure 6-3).

Atoms of metals in Groups IA-IIIA form cations as follows:

$$M\cdot \longrightarrow M^{+} + e^{-}$$
$$M: \longrightarrow M^{2+} + 2e^{-}$$
$$\overset{\cdot}{M}: \longrightarrow M^{3+} + 3e^{-}$$

The cation has the same number of positive charges as its group number.

Atoms of nonmetals in Groups VA-VIIA form anions as follows:

$$\cdot\overset{\cdot}{Z}: + 3e^{-} \longrightarrow :\overset{\cdot\cdot}{Z}: \;\; ^{3-}$$
$$\cdot\overset{\cdot\cdot}{Z}: + 2e^{-} \longrightarrow :\overset{\cdot\cdot}{Z}: \;\; ^{2-}$$
$$:\overset{\cdot\cdot}{Z}: + 1e^{-} \longrightarrow :\overset{\cdot\cdot}{Z}: \;\; ^{-}$$

Atoms of Group IVA elements have a similar tendency to gain or lose four electrons.  They seldom do either and are not often found as ions.

In the formation of ionic compounds, the total number of electrons lost by all metal atoms must be the same as the total number of electrons gained by all nonmetal atoms.

Ionic compounds have equal positive and negative charges and are electrically neutral.

## 6.4  Ionic Compounds

Ionic compounds exist in the solid state as crystals.  The anions and cations of a crystal are arranged in an orderly array called a crystal lattice (Figure 6-4).

Ionic bonds are not localized to a specific pair of ions but extend throughout the crystal.

A formula unit represents the simplest ratio of ions needed for electrical neutrality.

Attractive forces between cations and anions are strong.  It
is difficult to separate ions.  Consequently, ionic compounds
have high melting and boiling points and tend to be solids at
room temperature (Table 6-1).

6.5    An Introduction to Oxidation and Reduction

For uncombined atoms and monatomic ions, the oxidation number
of the atom or ion has the same value as the charge.  When
oxidation numbers are assigned to atoms or ions, the sign is
written in front of the number.

In oxidation-reduction reactions, one substance is oxidized
and another is reduced.  Oxidation is a loss of electrons or
an increase in oxidation number.  Reduction is a gain of
electrons or a decrease in oxidation number.

6.6    Covalent Bonds

Molecular compounds contain nonmetals held together in
molecules by covalent bonds.

A covalent bond is composed of a pair of electrons shared by
the nuclei of two atoms (Figures 6-7 and 6-8).

The molecular compound is stable when covalent bonds result
in each atom achieving a noble gas configuration.

Valence electrons not involved in bonding are called nonbonding electrons, unshared pairs, or lone pairs.

Seven elements exist in their uncombined states as diatomic
molecules (Table 6-2).

6.7    Molecular Compounds

Nonmetals have similar tendencies to gain or lose electrons,
so they share electrons with each other.

Molecular compounds can be solids, liquids or gases at room
temperature.

Attractive forces among molecules in covalent compounds are
relatively weak.  This accounts for their low melting and
boiling points.

6.8    Multiple Covalent Bonds

A double covalent bond occurs when two pairs of electrons are
shared by two atoms.

When three pairs of electrons are shared between two atoms,
the bond is called a triple bond.

6.9    Polyatomic Ions and Coordinate Covalent Bonds

An ion containing two or more atoms is called a polyatomic
ion.  The atoms are covalently bonded together.

Polyatomic ions may have positive or negative charges.  The charges arise from an excess or deficiency of electrons relative to protons in the ion.

A covalent bond composed of a pair of electrons provided by only one of the bonded atoms, is called a coordinate covalent bond.

6.10    Electron Dot Formulas of Molecular Compounds and Polyatomic Ions.

Valence is the number of covalent bonds an atom can form.

Electron dot formulas are based on chemical formulas and common valences of the atoms in the formula.

Common valences for several important elements are:  carbon, 4; nitrogen, 3; oxygen, 2; and hydrogen, 1.

To write electron dot formulas,
1.  Select the central atom and draw the atomic skeleton with single bonds.
2.  Count total valence electrons.
3.  Find the number of nonbonding electrons and distribute them.
4.  Check to see that each atom has a noble gas valence shell configuration.  Create multiple bonds if necessary.

6.11    Molecular Shapes.

Bonding and nonbonding electrons occur in pairs around atoms of molecular compounds.  Each valence shell electron pair repels other nearby pairs due to their negative charges.

The valence shell electron pair repulsion (VSEPR) theory predicts bond angles and molecular shapes based on the number of valence shell electron pairs around the central atom (Figures 6-10, 12, 13).

Formulas that indicate the geometry of molecules are called structural formulas (Figure 6-14).

6.12    Electronegativity.

Electronegativity is a measure of the attraction that a bonded atom has for the electrons in a bond.

Linus Pauling developed relative electronegativity values for the elements based on the most electronegative element, fluorine.

Electronegativity increases within a period as atomic size decreases, and decreases within a group as atomic size increases (Figure 6-15).

## 6.13  Polar Covalent Bonds

When atoms of different elements are covalently bonded, their different electronegativities result in an unequal attraction for the bonded electrons.  The electrons are shifted toward the more electronegative element.  The bond is called a polar covalent bond.

Differences in electronegativity between bonded atoms can be used to predict the nature of the bond.  If the differences are:

1.  less than 0.5, the bond is nonpolar covalent.
2.  greater than 2.0, the bond is ionic.
3.  0.5 - 2.0, the bond is polar covalent.

## 6.14  Polarity of Molecules

A polar molecule has an unsymmetrical distribution of electric charge.

Binary molecules with polar bonds are polar molecules.

For molecules containing three or more atoms and at least one polar covalent bond, symmetrical molecules will be nonpolar and unsymmetrical molecules will be polar.

---

SOLUTIONS TO SELECTED TEXT STUDY QUESTIONS AND PROBLEMS
and
PRACTICE PROBLEMS

---

11.  Give the formula of the ionic compound composed of each pair of elements.

a.  Cs and O        c.  Ca and At        e.  Na and N
b.  Al and S        d.  Ba and Se        f.  K and Se

The following steps summarize the method used in finding formulas of ionic compounds.

1.  From the periodic table, determine the group number for each element.
2.  Write the electron dot formula for each element.  The number of dots shown will be the same as the group number for all A group elements.  Paired electrons must be shown as paired dots.
3.  Predict the ion that will form for each element.  Cations form when metals lose all of their valence electrons.  Therefore, cations will have a positive charge equal to their group number.  Anions form when nonmetals gain electrons to achieve eight electrons in their valence shell.  Anions will have a negative charge equal to (group number - 8).
4.  Write the electron dot formula for each ion, showing the charge as a number followed by + or - as a super-script.  Note that cations have no dots since all

valence electrons are lost.  Think of the element's
symbol in the electron dot formula as the nucleus and
all inner energy level electrons.  Only valence shell
electrons are shown as dots.  The cation has an octet
of electrons remaining in the previous energy level,
and is therefore stable.

5.  Write the electron dot formulas of the cation and
anion together choosing appropriate subscripts to make
the total positive charge equal to the total negative
charge.  For clarity, parenthesis can be used around
ions with subscripts of 2 or more.

After a little practice, several steps in the above procedure
can be condensed.  Electron dot formulas for ions can be
written by examining the periodic table.

a.  Cs and O

1.  Cs is a Group IA metal, O is a Group VI A nonmetal.
2.  $Cs\cdot$, $\cdot\ddot{O}:$  Position of the dots is not important as
    long as they show correct number of paired and un-
    paired electrons.  Since oxygen has a valence
    electron configuration of $2s^2 2p^4$, 2s electrons are
    paired, and 2p electrons are paired.
3.  $Cs\cdot$ loses $1e^-$, to empty the valence energy level and
    achieve an octet of electrons in the previous energy
    level.  This results in a $1^+$ cation, which is
    expected for a Group I metal.  $\cdot\ddot{O}:$ gains $2e^-$, to
    complete the valence shell octet.  Thus results in a
    $2^-$ anion, which is expected for a Group VI nonmetal.
    (Group number $-8$) = $(6-8)$ = $-2$.
4.  $Cs^+$, $:\ddot{O}:^{2-}$  Notice that the sign of the ion follows
    the number.
5.  $(Cs^+)_2 :\ddot{O}:^{2-}$  Notice that the total positive charge
    $(+2)$ equals the total negative charge $(-2)$.

b.  Al and S

1.  Al is a Group IIIA metal, S is a Group VIA nonmetal.
2.  $Al:$ , $\cdot\ddot{S}:$
3.  $Al:$ loses $3e^-$, resulting in a $3^+$ cation, which is
    expected for a Group III metal.  $\cdot\ddot{S}:$ gains $2e^-$,
    resulting in a $2^-$ anion, which is expected for a
    Group VI nonmetal.  (Group number $-8$) = $(6-8)$ = $-2$.
4.  $Al^{3+}$, $:\ddot{S}:2-$
5.  $(Al^{3+})_2 (:\ddot{S}:^{2-})_3$  The total positive charge $(+6)$
    equals the total negative charge $(-6)$.

c.  Ca and At

Ca is a Group IIA metal.  Ca: loses $2e^-$, forming $Ca^{2+}$.
At is a Group VIIA nonmetal.  $\cdot\ddot{A}t:$ gains $1e^-$, forming $:\ddot{A}t:^-$.
$Ca^{2+} (:\ddot{A}t:^-)_2$  The total positive charge $(+2)$ equals the
total negative charge $(-2)$.

d.  Ba and Se

Ba is a Group IIA metal.  Ba: loses $2e^-$, forming $Ba^{2+}$.
Se is a Group VIA nonmetal.  ·Sе: gains $2e^-$ forming
:Sе:$^{2-}$.
$Ba^{2+}$ :Sе:$^{2-}$   The total positive charge (+2) equals
the total negative charge (-2).

e.  Na and N

Na is a Group IA metal.  Na· loses $1e^-$, forming $Na^+$.
N is a Group VA nonmetal.  ·N· gains $3e^-$, forming
:N:$^{3-}$.
$(Na^+)_3$ :N:$^{3-}$   The total positive charge (+3) equals
the total negative charge (-3).

f.  K and Se

K is a Group IA metal.  K· loses $1e^-$, forming $K^+$.
Se is a Group VIA nonmetal.  ·Sе: gains $2e^-$, forming
:Sе:$^{2-}$.
$(K^+)_2$ :Sе:$^{2-}$   The total positive charge (+2) equals
the total negative charge (-2).

***Practice Problem A

Fill in the table below to determine the formulas of several
ionic compounds.

| | element | $e^-$ dot for atom | $e^-$ dot for ion | formula |
|---|---|---|---|---|
| example | Li | Li· | $Li^+$ | $(Li^+)_3$ :P:$^{3-}$ |
| | P | ·P· | :P:$^{3-}$ | |
| | Ga | Ga: | $Ga^{3+}$ | $2Ga_3^{3+}$ :O:$^{2-}$    $Ga_2^3 O_3^2$ |
| | O | :O: | :O:$^{2-}$ | |
| | Sr | Sr: | $Sr_3^{2+}$ | $Sr_2^{3} S^{2-}$  $SrS$ |
| | S | :S: | :S:$^{2-}$ | |
| | Be | Be: | $Be^{2+}$ | $Be^{2+} Cl_2^-$ |
| | Cl | :Cl: | :Cl:$^-$ | |

34.    Write the electron dot formula of each of the following
       compounds.

       a.  $CaSO_4$          c.  $K_3PO_4$         e.  $Mg(C_2H_3O_2)_2$
       b.  $NaNO_3$          d.  $Na_2SO_4$        f.  $Sr(ClO_4)_2$

Each of the above compounds contains a polyatomic anion and
a metal cation.  Determine the charge of each cation.  From
the total positive charge, find the charge of the polyatomic
anion.

The atoms in polyatomic ions are covalently bonded together.
Follow the guidelines given in section 6.10 of the Summary
Outline to determine the e⁻ dot formula of the polyatomic
ion.

a.  $CaSO_4$

    Ca is a Group IIA metal.  Ca: loses $2e^-$, forming $Ca^{2+}$.

    Since the total positive charge is 2, the sulfate anion
    has a 2⁻ charge.

    1)  In $XO_n$ compounds or ions, X is the central atom.
        Arrange the 4 oxygen atoms around sulfur with single
        bonds.

$$\left[\begin{array}{c} O \\ | \\ O-S-O \\ | \\ O \end{array}\right]^{2-}$$

        The brackets are used to illustrate that the 2⁻
        charge is associated with the whole ion.

    2)  The total number of valence electrons are

            Valence e⁻ in a S atom            6
            Valence e⁻ in 4 O atoms          24
            Plus 2e⁻, as indicated by charge   2
            Total number of valence e⁻        32

    3)  8 electrons are already shown in the 4 single bonds
        in the skeleton above.  32−8=24 remaining electrons
        to be distributed as nonbonding electrons.  This gives

$$\left[\begin{array}{c} :\ddot{O}: \\ | \\ :\ddot{O}-S-\ddot{O}: \\ | \\ :\ddot{O}: \end{array}\right]^{2-}$$

    4)  Each element has 8 valence shell electrons, so the
        electron dot formula for sulfate is correct.  No
        multiple bonds are necessary.

        Combine the electron dot formulas of the cation and
        anion.

$$Ca^{2+} \quad \left[\begin{array}{c} :\ddot{O}: \\ | \\ :\ddot{O}-S-\ddot{O}: \\ | \\ :\ddot{O}: \end{array}\right]^{2-}$$

c.   $K_3PO_4$

K is a Group IA metal.  K· loses 1e⁻, forming $K^+$.  3 $K^+$
cations results in a total positive charge of (+3).

The total negative charge must be (−3).  One phosphate
ion accounts for this total charge, so the phosphate ion
carries a (−3) charge.

1)   The atomic skeleton is

$$\left[\begin{array}{c} O \\ | \\ O-P-O \\ | \\ O \end{array}\right]^{3-}$$

2)   The total valence electrons are

| | |
|---|---|
| 1 P atom | 5 |
| 4 O atoms | 24 |
| 3 e⁻ in excess | 3 |
| Total valence e⁻ | 32 |

3)   The number of nonbonding electrons are

32 valence e⁻ −8 bonding e⁻ = 24 nonbonding e⁻

$$\left[\begin{array}{c} :\ddot{O}: \\ | \\ :\ddot{O}-P-\ddot{O}: \\ | \\ :\ddot{O}: \end{array}\right]^{3-}$$

4)   Each element has 8 valence shell electrons.  No
multiple bonds are necessary.

Combine the electron dot formulas of the cation and
aion.

$$(K^+)_3 \quad \left[\begin{array}{c} :\ddot{O}: \\ | \\ :\ddot{O}-P-\ddot{O}: \\ | \\ :\ddot{O}: \end{array}\right]^{3-}$$

f.   $Sr(ClO_4)_2$

Sr is a Group IIA metal.  Sr: loses 2e⁻, forming $Sr^{2+}$.

The total negative charge must be (−2).  Two chlorate
ions account for this total charge, so each chlorate
ion carries a (−1) charge.

1)   The atomic skeleton is

$$\left[\begin{array}{c} O \\ | \\ O-Cl-O \\ | \\ O \end{array}\right]^{-}$$

2)  The total valence electrons are

| 1 Cl atom | 7 |
| 4 O atoms | 24 |
| 1 $e^-$ in excess | 1 |
| Total valence $e^-$ | 32 |

3)  The number of nonbonding electrons are

32 valence $e^-$ -8 bonding $e^-$ = 24 nonbonding $e^-$

$$\left[ \ddot{O}\text{-Cl-}\ddot{O} \right]^-$$

4)  Each element has 8 valence shell electrons.  No multiple bonds are necessary.

Combine the electron dot formulas of the cation and anion.

$$Sr^{2+} \quad \left[ \ddot{O}\text{-Cl-}\ddot{O} \right]^-_2$$

Consult Appendix C in the text for answers to parts b, d, and e.

***Practice Problem B

Write the electron dot formula of each of the following compounds.

a.  $CS_2$

b.  $NH_4ClO_3$

c.  $Rb_2O$

38.    Predict the shape and bond angles and give the structural
       formula of each of the following molecules.

   a.  $H_2S$              b.  $PCl_3$              c.  $CF_4$

   The following steps summarize a method for determining
   molecular geometry by the VSEPR theory.

   1)  Write the electron dot formula for the molecule.
   2)  Determine the number of "sets" of electron pairs
       about the central atom.  To count sets of electron
       pairs, examine the following examples.

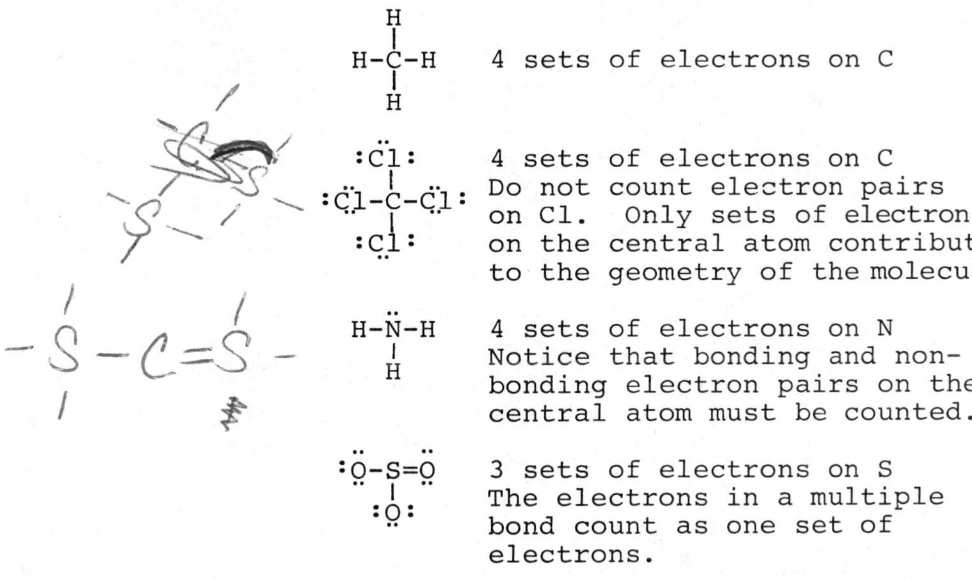

H–C–H    4 sets of electrons on C

:Cl–C–Cl:    4 sets of electrons on C
             Do not count electron pairs
             on Cl.  Only sets of electrons
             on the central atom contribute
             to the geometry of the molecule.

H–N–H    4 sets of electrons on N
         Notice that bonding and non-
         bonding electron pairs on the
         central atom must be counted.

:O–S=O    3 sets of electrons on S
          The electrons in a multiple
          bond count as one set of
          electrons.

   3)  Determine the geometry of the molecule based on the
       number of sets of electron pairs and the number of
       pairs of nonbonding electrons on the central atom.
       The geometry will be one that minimizes repulsions
       between the sets of electrons on the central atom.
       (Refer to Figures 6-10, 11, 12, 13 and 14 in the text).
       The following tables summarize these geometries.

| Number of Sets on the Central atom | Number of Nonbonding e⁻ pairs | Geometry | | Bond angles |
|---|---|---|---|---|
| 4 | 0 | | tetrahedral | 109.5° |
|   | 1 | | pyramidal | <109.5° |
|   | 2 | | bent | <109.5° |

Notice that 4 sets of electron pairs have electron repulsions minimized by a tetrahedral arrangement. The X's shown on the drawings locate a position for nonbonding electrons. These electrons occupy a region of space and contribute to the overall arrangement of the molecule. However only atoms connected by bonded electrons are considered in the shape of the molecule. The lines with X's are not considered in the geometry of the molecule.

The wedges are shown in the drawings to illustrate clearly the three dimensional nature of the molecule. The wedges indicate bonds in front or in back of the plane of the paper. The solid lines indicate bonds in the plane of the paper.

| Number of Sets on the Central atom | Number of Nonbonding e⁻ pairs | Geometry | Bond angles |
|---|---|---|---|
| 3 | 0 | trigonal | 120° |
|   | 1 | bent | <120° |
| 2 | 0 | linear | 180° |

The tables above do not consider cases where only one bond is present. These molecules all have linear geometry.

a.  $H_2S$

1)  H–S̈–H
2)  Sulfur has 4 sets of electron pairs.
3)  The 4 sets on sulfur can be drawn as,

then "X-out" the 2 nonbonding pairs.

The remaining lines indicate a bent molecule. The bond angles are those of a tetrahedron reduced by the greater repulsion of the nonbonding electrons.

The structural formula is ⟶
the shape is bent, and the
bond angles are less than
109.5°.

b.  $PCl_3$

    1)  :C̈L–P̈–C̈L:
           |
          :C̈l:

    2)  Phosphorus has 4 sets of electron pairs.
    3)  The 4 sets on phosphorus can be drawn as,

then "X-out" the one nonbonding pair.

The remaining lines indicate a pyramidal molecule.

The structural formula is ⟶
the shape is pyramidal, and
the bond angles are less than
109.5°.

Consult Appendix C in the text for the answer to part c.

***Practice Problem C

Predict the shape and bond angles and give the structural formula of the following molecules or ions.

a.  $SO_2$

b.  $Cl_2O$

47.  Classify each of the following molecules as polar or non-polar and explain your answer.

a.  $CBr_4$
b.  $H_2S$
c.  $Br_2$

Molecular polarity depends on the polarity of the bonds within the molecule and the geometry of the molecule.  The following steps summarize a method for determining molecular polarity.

1) Write the electron dot formula for the molecule.

2) Determine the molecular geometry. Draw a structural formula.

3) Calculate the electronegativity difference in each bond. Label the partial charges that result from electronegativity differences.

4) Based on the shape of the molecule, determine whether the electrical charge is distributed symmetrically or unsymmetrically.

a. $CBr_4$

1)

```
        ..
      :Br:
        |
  ..    |    ..
 :Br-C-Br:
        |
      :Br:
        ..
```

2) The molecule is tetrahedral.

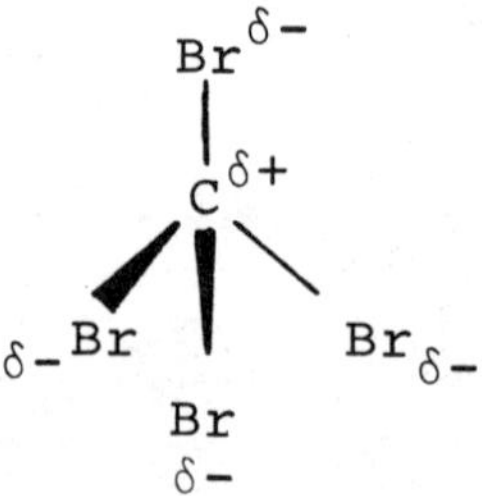

3) The electronegativity difference in a Br-C bond is

$$2.8-2.5 = 0.3$$

The electron withdrawing effects of Br are small. The bonds can be labeled.

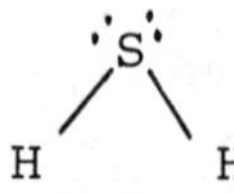

4) The bromine atoms withdraw bonding electrons with equal force in opposite directions, resulting in a symmetrical distribution of charge. The molecule is nonpolar.

b. $H_2S$

1)  H-S-H
2)  The molecule is bent.

3) The electronegativity difference in a S-H bond is

$$2.5-2.1 = 0.4$$

The bonds can be labeled

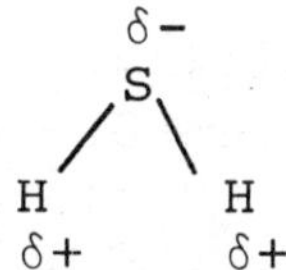

4) The sulfur atom withdraws bonding electrons from hydrogen atoms which are not symmetrically arranged. The electrical charge is distributed unsymmetrically. The molecule is polar.

Consult Appendix C in the text for the answer to part c.

***Practice Problem D

Classify each of the following molecules as polar or non-polar and explain your answer.

a. HCN *polar*

*because N e⁻ are decreasing and H has the least electronegativity.*

b. $PCl_3$ *polar*

*The Cl atoms are bonding w/ the phosphorus the electromagnetic charge is unequal.*

---

### SELF-TEST

---

Circle the correct answer in the following multiple choice questions.

1. Which of the following compounds has ionic bonds?

   a. $Cl_2O$
   b. $Na_2O$
   c. $H_2O$
   d. $F_2O$

2. Which of the following statements is true?

   a. $Ca^{2+}$ is larger than Ca
   b. $O^{2-}$ is smaller than O
   c. $Na^+$ is smaller than Na
   d. $F^-$ is smaller than F

3.  All elements of Group VIA

    a.  are metals.
    b.  form negative six (-6) ions.
    c.  have electron dot formulas with 6 dots, $\cdot \ddot{X} \colon$.
    d.  form cations.

4.  Which of the following is correct for the formation of an ion?

    a.  Al lose 3 electrons to form $Al^{3-}$.
    b.  S loses 2 electrons to form $S^{2-}$.
    c.  Mg gains 2 electrons to form $Ca^{2+}$.
    d.  N gains 3 electrons to form $N^{3-}$.

5.  For the reaction, $Al + Au^{3+} \longrightarrow Au + Al^{3+}$

    a.  Al is being oxidized.
    b.  Au is being oxidized.
    c.  $Au^{3+}$ is losing electrons.
    d.  Al is gaining electrons.

6.  How many nonbonding electrons are present on sulfur in $SO_2$?

    a.  0
    b.  1
    c.  2
    d.  4

7.  How many valence electrons are in $SO_4{}^{2-}$?

    a.  28
    b.  30
    c.  32
    d.  34

8.  Which of the following electron dot formulas is correct?

    a.  $Na-\ddot{C}l\colon$

    b.  $H-\ddot{O}=H$

    c.  $\colon C \equiv O \colon$

    d.  $\colon \ddot{O}-\ddot{S}-\ddot{O} \colon$

9.  The shape of the $ClO_2{}^-$ ion is

    a.  linear
    b.  bent
    c.  pyramidal
    d.  tetrahedral

10. Which of the following linear molecules is nonpolar?

    a.  $CO_2$
    b.  $CO$
    c.  HBr
    d.  HCN

---

## ANSWERS TO PRACTICE PROBLEMS

### Practice Problem A

| element | e⁻ dot for atom | e⁻ dot for ion | formula |
|---------|-----------------|----------------|---------|
| Ga | Ga: | $Ga^{3+}$ | $Ga_2O_3$ |
| O | ·Ö: | $:\ddot{O}:^{2-}$ | |
| Sr | Sr: | $Sr^{2+}$ | SrS |
| S | ·S̈: | $:\ddot{S}:^{2-}$ | |
| Be | Be: | $Be^{2+}$ | $BeCl_2$ |
| Cl | ·C̈l: | $:\ddot{C}l:^{-}$ | |

### Practice Problem B

a.   $CS_2$

Since both C and S are nonmetals, $CS_2$ is a molecular compound.  The atomic skeleton is S–C–S.  The total valence electrons are

    1 C atom             4
    2 S atoms           12
    Total valence e⁻    16

The number of nonbonding electrons are

$$\left(16 \text{ valence } e^-\right) - \left(4 \text{ bonding } e^-\right) = 12 \text{ nonbonding } e^-$$

$$:\ddot{S}-C-\ddot{S}: \longrightarrow \ddot{S}=C=\ddot{S}$$

b.   $NH_4ClO_3$

Both $NH_4^+$ and $ClO_3^-$ are polyatomic ions with covalent bonds.  $NH_4^+$ has 8 valence electrons.

    1 N atom             5
    4 H atoms            4
    less 1 electron     -1
    Total valence e⁻     8

$ClO_3^-$ has 26 valence electrons.

    1 Cl atom            7
    3 O atoms           18
    1 excess electron    1
    Total valence e⁻    26

The bonds skeletons are

$$\left[ \begin{array}{c} H \\ | \\ H-N-H \\ | \\ H \end{array} \right]^{+} \qquad \text{and} \qquad \left[ \begin{array}{c} O-Cl-O \\ | \\ O \end{array} \right]^{-}$$

$NH_4^+$ is complete. $ClO_3^-$ has 20 nonbonding electrons.

$(26 \text{ valence } e^-) - (6 \text{ bonding } e^-) = 20 \text{ nonbonding } e^-$

$$\left[ :\ddot{O}-\ddot{C}l-\ddot{O}: \atop :\ddot{O}: \right]^{-}$$

Each element in $ClO_3^-$ has 8 valence shell electrons.
Combine the electron dot formulas of the cation and aion.

$$\left[ \begin{array}{c} H \\ | \\ H-N-H \\ | \\ H \end{array} \right]^{+} \qquad \left[ :\ddot{O}-\ddot{C}l-\ddot{O}: \atop :\ddot{O}: \right]^{-}$$

c.   $Rb_2O$

Rb is a metal and O is a nonmetal, so $Rb_2O$ is an ionic compound.
Rb is a group IA metal. $Rb\cdot$ loses 1 $e^-$, forming $Rb^+$.
O is a group VIA nonmetal. $\cdot\ddot{O}:$ gains $2e^-$, forming $:\ddot{O}:^{2-}$.

$$(Rb^+)_2 \quad :\ddot{O}:^{2-}$$

The total positive charge (+2) equals the total negative charge (−2).

Practice Problem C

a.   $SO_2$

1)  $\ddot{O}=\ddot{S}-\ddot{O}:$
2)  S has 3 sets of electron pairs.
3)  The 3 sets on S can be drawn as,

then "X-out" the one nonbonding pair.

The remaining lines indicate a bent molecule.

The structural formula is

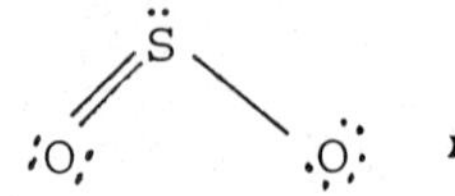

the shape is bent, and the bond angle is less than 120°.

b.  $Cl_2O$

   1)  $:\!\ddot{C}l\!-\!\ddot{O}\!-\!\ddot{C}l\!:$

   2)  O has 4 sets of electron pairs.

   3)  The 4 sets on O can be drawn as,

then "X-out" the 2 nonbonding pairs.

The remaining lines indicate a bent molecule.

The structural formula is

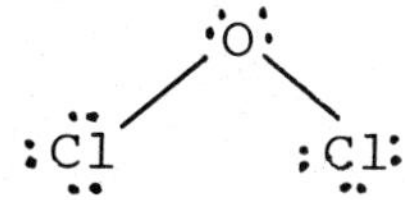

the shape is bent, and the bond angle is less than 109.5°.

## Practice Problem D

a.  HCN

   1)  $H\!-\!C\!\equiv\!N\!:$

   2)  The molecule is linear.

   3)  The electronegativity difference in each bond is

        C–H = 2.5-2.1 = 0.4
        N–C = 3.0-2.5 = 0.5

   4)  Since N has the largest electron withdrawing effect and H has the smallest, the molecule is polar

$$\overset{\longrightarrow}{\underset{H-C\equiv N:}{\vdash}}$$

b.  $PCl_3$

   1)  $:\!\ddot{C}l\!-\!\ddot{P}\!-\!\ddot{C}l\!:$
           $|$
         $:\!\ddot{C}l\!:$

   2)  The molecule is pyramidal

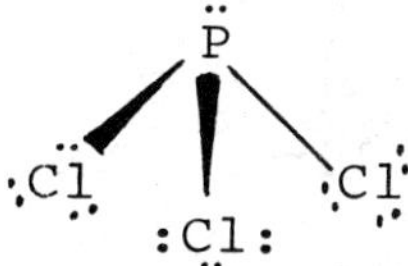

   3)  The electronegativity difference in the Cl-P bond is

        3.0-2.1 = 0.9

The bonds can be labeled

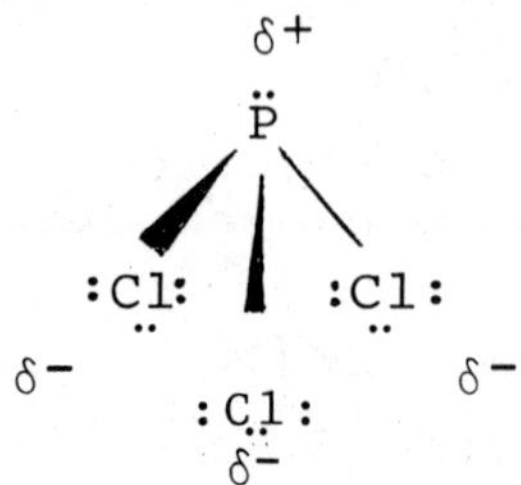


4)   The chlorine atoms withdraw bonding electrons from the phosphorus atom.  The electrical charge is distributed unsymmetrically.  $PCl_3$ is polar.

---

## ANSWERS TO SELF-TEST

1.  b
2.  c
3.  c
4.  d
5.  a
6.  c
7.  c
8.  c
9.  b
10.  a

# NAMING INORGANIC COMPOUNDS

# CHAPTER 7

CHAPTER 7
NAMING INORGANIC
COMPOUNDS

---

## SUMMARY OUTLINE

### 7.1 Oxidation Numbers

An oxidation number is a number assigned to an element that is uncombined or in a compound.

Oxidation numbers can be zeroes or signed numbers.

Oxidation numbers are useful as mechanical aids in writing formulas, naming inorganic compounds and balancing equations.

The following rules summarize the assignment of oxidation numbers to elements:

1. Uncombined elements have an oxidation number of zero.
2. The oxidation number of a monatomic ion is the same as its charge.
3. Hydrogen in compounds has an oxidation number of +1, except when combined with a less electronegative element. Then hydrogen has an oxidation number of -1.
4. Oxygen in compounds has an oxidation number of -2, except in peroxide compounds. Then oxygen has an oxidation number of -1.
5. The sum of the oxidation numbers in a compound is zero. In a polyatomic ion, the sum of the oxidation numbers is equal to the charge of the ion.

### 7.2 Binary Ionic Compounds

To name a binary ionic compound, write the name of the metallic element first, followed by a separate word containing the root of the nonmetallic element name and the suffix -ide (Table 7-2).

If a metal cation has several possible charges, its oxidation number must be specified using Roman numerals in parentheses immediately following the name of the metal.  An alternate method names the metal using the root of the name from which the element's symbol is derived followed by the suffix -ous or -ic.  The ending -ous is used for the cation of lower oxidation state, and the ending -ic is used for the cation of higher oxidation state (Table 7-3).

7.3    Ionic Compounds Containing Polyatomic Ions

To name ionic compounds containing polyatomic ions, write the cation name first, followed by the anion name as a separate word (Table 7-4).

Polyatomic ions are written in parentheses with a corresponding subscript to indicate the presence of two or more of the same polyatomic ion in a formula unit.

Oxyanions are polyatomic ions containing oxygen and one other element.

Different oxyanions of the same element are named according to the oxidation state of the element.  When two oxyanions of the same element exist,
    the -ite ending is used for the lower oxidation state.
    the -ate ending is used for the higher oxidation state.

When four oxyanions of the same element exist,
    the hypo- prefix and -ite suffix are used for the lowest oxidation state,
    the hyper- prefix and -ate suffix are used for the highest oxidation state, and
    the -ite and -ate endings are used for the intermediate oxidation states.

7.4    Binary Molecular Compounds

The following rules are used to name binary molecular compounds

1.   Name the less electronegative element using the complete English name.
2.   Name the more electronegative element using the English root followed by the -ide ending.
3.   Use Greek prefixes to indicate the number of atoms of the element in the molecule (Table 7-5).
4.   The prefix mono- is not normally used with the less electronegative element.

7.5    Acids.

Acids are molecular compounds that produce $H^+$ when dissolved in water.

Acids usually contain hydrogen written as the first element.

To name binary acids, add the prefix <u>hydro-</u> and the suffix <u>-ic</u> to the root of the second element in the formula (Table 7-6).

Oxyacids contain an oxyanion and hydrogen atoms.

To name an oxyacid,
   change the <u>-ate</u> ending of the oxyanion to <u>-ic</u>, or
   change the <u>-ite</u> ending of the oxyanion to <u>-ous</u>, and
   add the word <u>acid</u>.

## 7.6    <u>Inorganic Hydrates</u>

Inorganic hydrates are ionic compounds that contain definite amounts of water in their crystals.

To name inorganic hydrates, write the name of the anhydrous compound followed by the word <u>hydrate</u>. Greek prefixes are used with the word <u>hydrate</u> to indicate the number of water molecules in the formula unit (Table 7-7).

---

SOLUTIONS TO SELECTED STUDY QUESTIONS AND PROBLEMS
and
PRACTICE PROBLEMS

---

4.    Find the oxidation number of each element in the following compounds.

   a.   $Rh_2(SO_4)_3$          c.   $C_{12}H_{22}O_{11}$          e.   $Cl_2O_7$
   b.   $H_4P_2O_5$             d.   $Po(IO_3)_4$

Using the rules for assigning oxidation numbers (section 7.1), write the oxidation number beneath each element with the sign of the charge in front of the number. In this way, oxidation numbers can be distinguished from ionic charges which are written as superscripts with the sign of the charge following the number.

   a.   $Rh_2(SO_4)_3$

   In this compound, only the oxidation number of oxygen is known. Both rhodium and sulfur must be deduced from the formula of the compound. This can be simplified if the sulfate ion is considered first. Table 7-4 gives the charge on the sulfate ion as 2-. From this charge, the oxidation number of sulfur in $SO_4^{2-}$ can be found.

$$\left(\begin{array}{c}\text{oxidation no.}\\\text{of S}\end{array}\right) + 4\left(\begin{array}{c}\text{oxidation no.}\\\text{of O}\end{array}\right) = \text{charge on the ion}$$

   (oxidation no. of S )+  4(-2)  = -2
   (oxidation no. of S)+  (-8)    = -2
   (oxidation no. of S )          = -2+8 = +6

Now find the oxidation number of rhodium.
In order for the formula unit to be neutral;
2(ionic charge of Rh) + 3 (charge of $SO_4^{2-}$) = o
To find the ionic charge of Rh, substitute the charge of
sulfate into the equation.
2(ionic charge of Rh) + 3 (-2) = 0
2(ionic charge of Rh) + (-6) = 0
2(ionic charge of Rh) = +6
 (ionic charge of Rh) = +3
The oxidation number of a monatomic ion is the same as
its charge, therefore the oxidation number of Rh is +3.
The oxidation numbers are $Rh_2 (SO_4)_3$.

$$+3 \quad +6-2$$

b.  $H_4P_2O_5$

The oxidation number of hydrogen is +1 and the oxidation
number of oxygen is -2.  The sum of the oxidation numbers
in the compound is zero.

$$4 \left( \begin{matrix} \text{oxidation no.} \\ \text{of H} \end{matrix} \right) + 2 \left( \begin{matrix} \text{oxidation no.} \\ \text{of P} \end{matrix} \right) + 5 \left( \begin{matrix} \text{oxidation no.} \\ \text{of O} \end{matrix} \right) = 0$$

$$4(+1) + 2 \left( \begin{matrix} \text{oxidation no.} \\ \text{of P} \end{matrix} \right) + 5 (-2) = 0$$

$$4 + 2 \left( \begin{matrix} \text{oxidation no.} \\ \text{of P} \end{matrix} \right) + (-10) = 0$$

$$2 \left( \begin{matrix} \text{oxidation no.} \\ \text{of P} \end{matrix} \right) + (-6) = 0$$

$$2 \left( \begin{matrix} \text{oxidation no.} \\ \text{of P} \end{matrix} \right) = +6$$

(oxidation no. of P) = +3

The oxidation of numbers are $H_4P_2O_5$.

$$+1+3-2$$

d.  $Po(IO_3)_4$

In this compound, only the oxidation number of oxygen is
known.  Both polonium and iodine must be deduced from the
formula of the compound.  Again, recognize that $IO_3$ is a
polyatomic ion and use Table 7-4 to find its total charge.
From this charge, determine the oxidation number of iodine:

$$\left( \begin{matrix} \text{oxidation no.} \\ \text{of I} \end{matrix} \right) + 3 \left( \begin{matrix} \text{oxidation no.} \\ \text{of O} \end{matrix} \right) = -1$$

$$\left( \begin{matrix} \text{oxidation no.} \\ \text{of I} \end{matrix} \right) + 3 (-2) = -1$$

$$\left( \begin{matrix} \text{oxidation no.} \\ \text{of I} \end{matrix} \right) + (-6) = -1$$

(oxidation no. of I) = +5

Now find the oxidation number of polonium.  Since,
(ionic charge of Po) + 4 (charge of $IO_3^-$) = 0
in order for the formula unit to be neutral; then,
(ionic charge of Po) + 4 (-1) = 0
(ionic charge of Po) = +4

The oxidation numbers are $Po(IO_3)_4$.
$$+4 \quad +5 \;-2$$

***Practice Problem A

Find the oxidation number of chlorine in each of the following compounds or ions.   (Answer in the space provided).

a.  $ClO^-$

$$(\text{ox. no. of } Cl) + (\text{ox. no. of } O) = -1$$
$$(\text{ox. no. of } Cl) + (-2) = -1$$
$$(\text{ox. no. of } Cl) = +1$$

b.  $ClO_2^-$

$$(\text{ox. no. of } Cl) + 2(\text{ox. no. of } O) = -1$$
$$(\text{ox. no. of } Cl) + 2(-2) = -1$$
$$(\text{ox. no. of } Cl) \; (-4) = -1$$
$$(\text{ox. no. of } Cl) = 3$$

c.  $ClO_3^-$

$$(\text{ox. no. } Cl) + 3(\text{ox. no. } O) = -1$$
$$(\text{ox. no. } Cl) + 3(-2) = -1$$
$$\text{ox. no. } Cl. \;\; -6 = -1$$
$$\text{ox. no. } Cl = +5$$

d.  $ClO_4^-$

$$(\text{ox. no. } Cl) + 4(\text{ox. no. } O) = -1$$
$$\text{ox. no. } Cl + 4(-2) = -1$$
$$\text{ox. no. } Cl = 7$$

e.  HCl

$$(\text{ox. no. of } H) + (\text{ox. no. } Cl) = 0$$
$$(H) + \text{ox. no. } Cl = 0$$
$$\text{ox. no. } Cl = -1$$

f.  $Cl_2O$

$$2(\text{ox. no. } Cl)(\text{ox. no. } O) = 0$$
$$2(\text{ox. no. } Cl) + (-2) = 0$$
$$2(\text{ox. no. } Cl) = 2$$
$$\text{ox. no. } Cl = +1$$

The charts shown on the following pages summarize the guidelines for naming compounds given in sections 7.2, 7.3, 7.4 and 7.5.

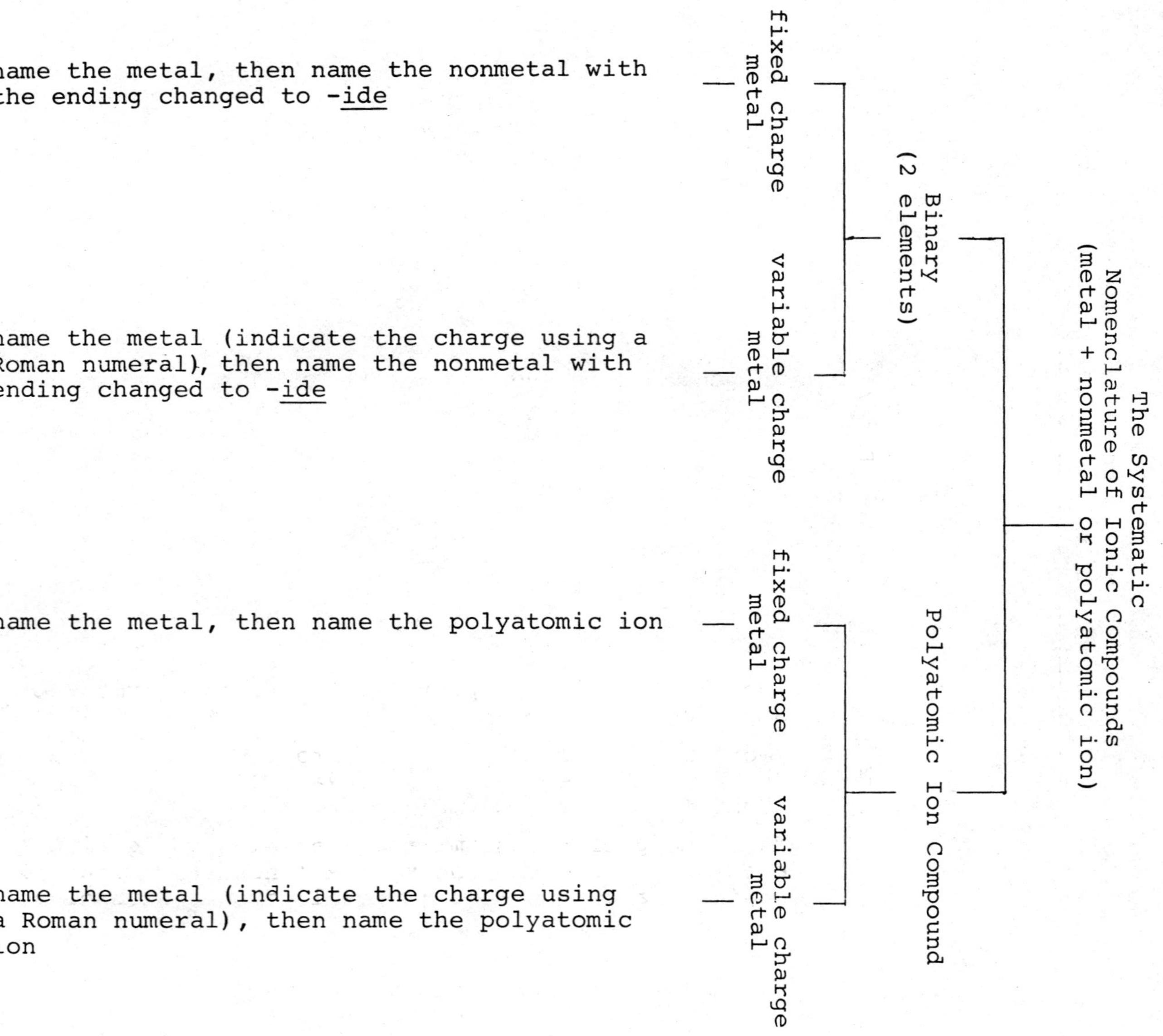

The Systematic
Nomenclature of Ionic Compounds
(metal + nonmetal or polyatomic ion)

Binary
(2 elements)

Polyatomic Ion Compound

fixed charge
metal

variable charge
metal

fixed charge
metal

variable charge
metal

name the metal, then name the nonmetal with the ending changed to -ide

name the metal (indicate the charge using a Roman numeral), then name the nonmetal with ending changed to -ide

name the metal, then name the polyatomic ion

name the metal (indicate the charge using a Roman numeral), then name the polyatomic ion

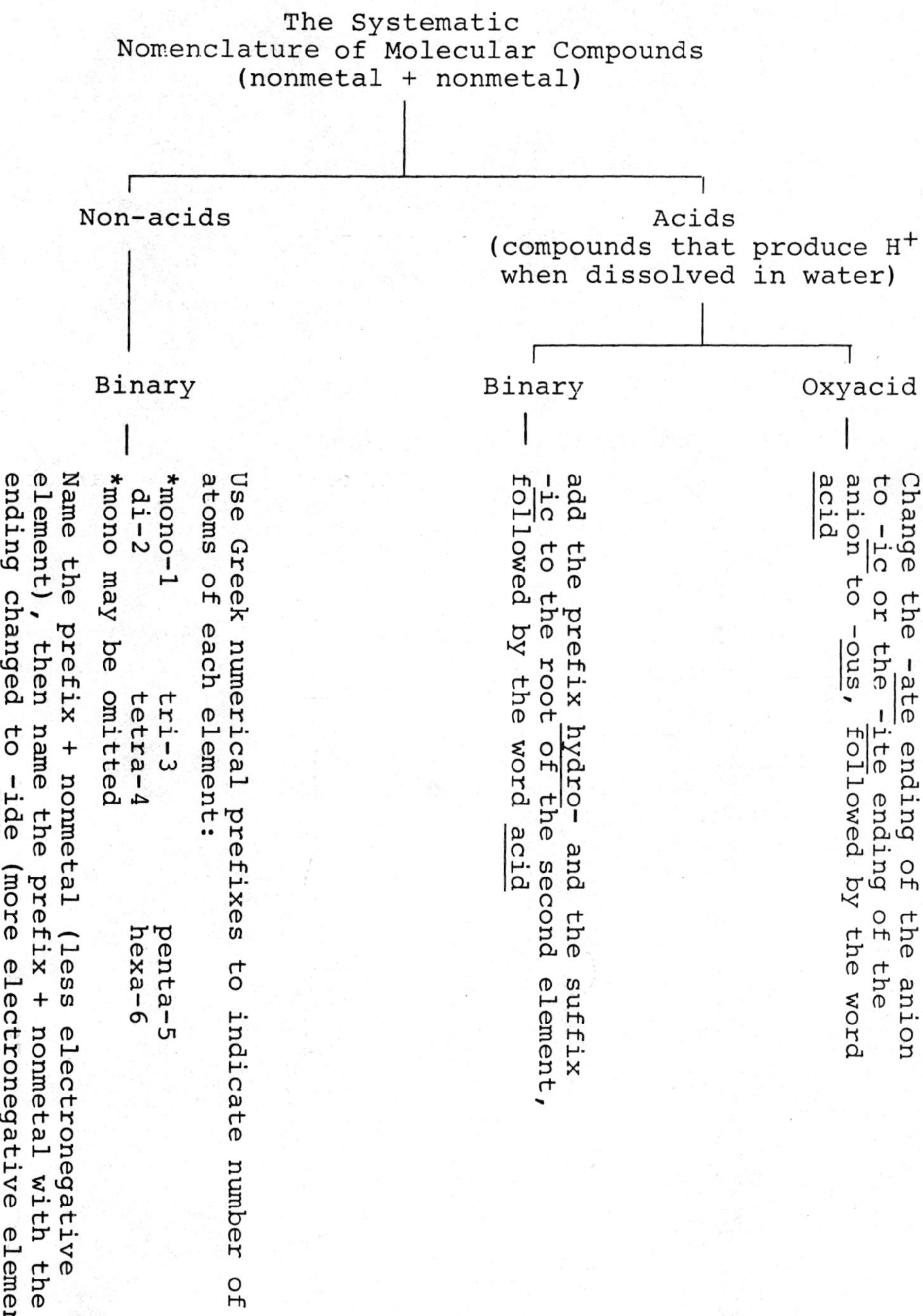
The Systematic
Nomenclature of Molecular Compounds
(nonmetal + nonmetal)

Non-acids

Acids
(compounds that produce H+
when dissolved in water)

Binary

Binary

Oxyacid

Use Greek numerical prefixes to indicate number of atoms of each element:
*mono-1     tri-3      penta-5
di-2        tetra-4    hexa-6
*mono may be omitted
Name the prefix + nonmetal (less electronegative element), then name the prefix + nonmetal with the ending changed to -ide (more electronegative element)

add the prefix hydro- and the suffix -ic to the root of the second element, followed by the word acid

Change the -ate ending of the anion to -ic or the -ite ending of the anion to -ous, followed by the word acid

5. Give the systematic name of each compound.

   The chart on page 94 summarizes the rules for naming binary ionic compounds.

   a. $GaI_3$, a binary ionic compound with a fixed charge metal, is named gallium iodide.

   b. FeS is a binary ionic compound with a variable charge metal. To determine the charge of Fe, first determine the charge of S. S is a Group VIA nonmetal, so it forms a $S^{2-}$ ion.
      ionic charge of Fe + ionic charge of S = 0
      ionic charge of Fe + (-2) = 0
      ionic charge of Fe = +2
      The name of FeS is iron(II) sulfide.

   Consult Appendix C in the text for answers to parts c and d

6. Write the formula of each compound.

   Write the formula and charge for each ion. Then combine the ions into a formula choosing appropriate subscripts to make the total positive charge equal to the total negative charge.

   a. Iron(II) iodide

      Iron(II) specifies the $Fe^{2+}$ ion. Iodine is a Group VIIA nonmetal, so it forms the $I^-$ ion. The formula is $FeI_2$.

   b. Sodium hydride

      Sodium, a Group IA metal, forms the $Na^+$ ion. Hydrogen forms the $H^-$ ion when combined with a less electronegative element. The formula is NaH.

   c. Zirconium(IV) oxide

      Zirconium(IV) specifies the $Zr^{4+}$ ion. Oxygen is a Group VIA nonmetal, so it forms the $O^{2-}$ ion. The formula is $ZrO_2$.

   d. Barium telluride

      Barium, a Group IIA metal, forms the $Ba^{2+}$ ion. Tellurium, a Group VIA nonmetal, forms the $Te^{2-}$ ion. The formula is BaTe.

   e. Mercury(II) fluoride

      Mercury(II) specifies the $Hg^{2+}$ ion. Fluorine, a Group VIIA metal, forms the $F^-$ ion. The formula is $HgF_2$.

10. Give the systematic name of each compound.

    The chart on page 94 summarizes the rules for naming ionic compounds containing polyatomic ions.

a.   $Al(HSO_4)_3$

Al has a fixed charge.  $HSO_4^-$ is named hydrogen sulfate or bisulfate (Table 7-4).  $Al(HSO_4)_3$ is named aluminum hydrogen sulfate or aluminum bisulfate.

c.   $Fe_2(HPO_4)_3$

Since Fe is a variable charge metal, its charge must be determined.  $HPO_4$ has a $2^-$ charge (Table 7-4).

2(ionic charge of Fe) + 3 (charge of $HPO_4$) = 0
2(ionic charge of Fe) + 3 (-2) = 0
2(ionic charge of Fe) = +6
  ionic charge of Fe  = +3

The name of $Fe_2(HPO_4)_3$ is iron(III) hydrogen phosphate or iron(III) biphosphate.

Consult Appendix C in the text for answers to parts b and d.

11.   Write the formula of each compound.

Write the formula and charge for each ion.  Then combine the ions into a formula choosing appropriate subscripts to make the total positive charge equal to the total negative charge.

a.   Barium hydroxide

Barium, a Group IIA metal, forms the $Ba^{2+}$ ion.  The hydroxide ion is $OH^-$ (Table 7-4).  Two $OH^-$ ions are needed to balance the $Ba^{2+}$ ion.  Parentheses must be used to indicate two $OH^-$ ions.  The formula is $Ba(OH)_2$.

b.   Potassium hydrogen sulfite

Potassium, a Group IA metal, forms the $K^+$ ion.  The hydrogen sulfite ion is $HSO_3^-$ (Table 7-4).  The formula is $KHSO_3$.

c.   Indium(III) perchlorate

Indium(III) specifies the $In^{3+}$ ion.  The perchlorate ion is $ClO_4^-$ (Table 7-4).  The formula is $In(ClO_4)_3$.

d.   Plumbous iodate

The plumbous ion has the lower oxidation state of the two lead ions, $Pb^{2+}$, and the iodate ion is $IO_4^-$ (Table 7-4). The formula is $Pb(IO_4)_2$.

e.   Cupric cyanide has the higher oxidation state of the two copper ions, $Cu^{2+}$, and the cyanide ion is $CN^-$ (Table 7-4). The formula is $Cu(CN)_2$.

12. Name each compound.

   a. $B_2Br_4$, a binary molecular compound, is named diboron tetrabromide.

   b. $BrF_3$, a binary molecular compound, is named bromine trifluoride.

   Consult Appendix C in the text for answers to parts c and d.

13. Write the formula of each compound.

   The formulas of molecular compounds can be determined directly from the name. Use the prefixes to write the subscripts.

   a. Xenon tetroxide

      If no prefix is written, <u>mono</u>- is understood, and one atom of the element is indicated. The formula is $XeO_4$.

   b. Tetrasulfur dinitride
      $S_4N_2$

   c. Hydrogen sulfide

      The names of binary molecular compounds containing hydrogen do not follow the guidelines given in section 7-4. No Greek numerical prefixes are used. Although no ions exist, the formula can be determined as if the compound is ionic. Sulfur, in Group VIA, forms $S^{2-}$ and hydrogen in Group IA forms $H^+$. The formula $H_2S$, determined in this way, corresponds to the formula of the molecule (Table 7-6).

   d. Disilicon hexaiodide
      $Si_2I_6$

   e. Diphosphorus tetrachloride
      $P_2Cl_4$

14. Name each acid.

   The chart on page 95 summarizes the systematic nomenclature of acids.

   b. $H_2SO_4$
      Oxyanion name: sulfate
      Acid name: sulfuric acid

c.   $HC_2H_3O_2$

Oxyanion name:  acetate
Acid name:  acetic acid

e.   HF

HF is a binary acid and is named hydrofluoric acid.

Consult Appendix C in the text for answers to parts a, and d.

15.   Write the formula of each compound.

a.   Sulfurous acid
b.   hydrosulfuric acid
c.   periodic acid
d.   oxalic acid
e.   chromic acid

Use the following guidelines to write formulas of acids.

1.   Determine if the compound is a binary acid or an oxyacid.  The prefix <u>hydro-</u> specifies a binary acid.
2.   Determine the charge on the nonmetal anion or polyatomic ion in the acid.  (The compound is a molecule. $H^+$ cations and nonmetal anions or polyatomic ions are formed when the molecule is added to water. The total negative charge in the acid is equal to the total positive charge from $H^+$ ions).

a.   Sulfurous acid

Sulfurous acid is an oxyacid containing the sulfite ion, $SO_3^{2-}$.  The formula is $H_2SO_3$.

b.   Hydrosulfuric acid

Hydrosulfuric acid is a binary acid containing the sulfide ion, $S^{2-}$.  The formula is $H_2S$.

c.   Periodic acid

Periodic acid is an oxyacid containing the periodate ion, $IO_4^-$.  The formula is $HIO_4$.

***Practice Problem B

Identify the compound as ionic or molecular.  Then name each compound.

a.   $Zn(OH)_2$      *ionic*      *zinc hydroxide*
b.   $CuI_2$         *ionic*      *copper (II) iodide*
c.   $P_2O_5$        *molecular*  *diphosphorus pentoxide*
d.   $HNO_2$         *molecular*  *nitrous acid*
e.   $Se_2Br_2$      *molecular*  *diselenium dibromide*
f.   $Pb(CO_3)_2$    *ionic*      *lead IV carbonate*
g.   $Li_3N$         *ionic*      *lithium nitrate*

|  |  | | |
|---|---|---|---|
| h. | $H_3P$ | _molecular_ | hydrophosphuric acid |
| i. | $CaCrO_4$ | _ionic_ | calcium cromate |
| j. | $AsCl_5$ | _molecular_ | arsenic pentachloride |

*** Practice Problem C

Write the formula of each compound.

| | | |
|---|---|---|
| a. | xenon tetrafluoride | $XeF_4$ |
| b. | magnesium chlorate | $Mg(ClO_3)_2$ |
| c. | hydroselenic acid | $H_2Se$ |
| d. | nickel(II) sulfide | $NiS$ |
| e. | dinitrogen trioxide | $N_2O_3$ |
| f. | hypobromous acid | $HBrO$ |
| g. | copper(I) sulfate | $Cu_2SO_4$ |
| h. | ammonium nitrate | $NH_4NO_3$ |
| i. | phosphorous acid | $H_3PO_3$ |
| j. | aluminum oxide | $Al_2O_3$ |

---

## SELF-TEST

Circle the correct answer in the following multiple choice questions.

1. The oxidation number for nitrogen is +3 in

    a. $HNO_2$
    b. $NH_3$
    c. $HNO_3$
    d. $NO_2$

2. The -__ic__ ending for iron in ferric chloride indicates that

    a. iron has the lower of two oxidation states.
    b. iron has a +1 oxidation state.
    c. iron has a +3 oxidation state.
    d. the compound is a binary acid.

3. The Roman numeral in tin(IV) hydroxide indicates that

    a. the formula is $Sn_4OH$.
    b. the oxidation state of tin is +4.
    c. the formula is $Sn(OH)_4$.
    d. the charge of tin is -4.

4. The __di__- prefix in diphosphorus pentoxide indicates that

    a. phosphorus has an oxidation state of +2.
    b. hydrogen is present in the polyatomic ion, $HPO_5^{3-}$.
    c. there are two phosphorus atoms in the molecule.
    d. the formula of the compound is $P_5O_2$.

5.   Which of the following is a binary acid?

    a.   chloric acid
    b.   hypochlorous acid
    c.   perchloric acid
    d.   hydrochloric acid

6.   Which of the following is not true for $CuSO_4 \cdot 5H_2O$?

    a.   Five molecules of water are present in each formula unit of $CuSO_4$ crystals.
    b.   The name of the compound is copper sulfate pentahydrate.
    c.   The composition of $CuSO_4 \cdot 5H_2O$ can vary.
    d.   The formula unit contains a total of nine atoms of oxygen.

7.   The systematic name for $CaF_2$ is

    a.   calcium difluoride.
    b.   calcium fluoride.
    c.   calcium(II) fluoride.
    d.   monocalcium difluoride.

8.   The correct formula for lead(IV) sulfate is

    a.   $Pb_2(SO_4)_4$
    b.   $Pb(SO_3)_2$
    c.   $Pb_4(SO_3)_2$
    d.   $Pb(SO_4)_2$

9.   The correct formula for nitrous acid is

    a.   $HNO_3$
    b.   $H_2NO_2$
    c.   $HNO_2$
    d.   $H_2NO_3$

10.   The correct formula for dinitrogen oxide is

    a.   $N_2O_2$
    b.   $2NO$
    c.   $NO_2$
    d.   $N_2O$

---

## ANSWERS TO PRACTICE PROBLEMS

**Practice Problem A**

    a.   $ClO^-$

$$(\text{oxidation no. of Cl}) + (\text{oxidation no. of O}) = -1$$
$$(\text{oxidation no. of Cl}) + (-2) = -1$$
$$(\text{oxidation no. of Cl}) = +1$$

b. $ClO_2^-$

(oxidation no. of Cl) + 2(oxidation no. of O) = -1
(oxidation no. of Cl) + 2 (-2) = -1
(oxidation no. of Cl) + (-4) = -1
(oxidation no. of Cl) = +3

c. $ClO_3^-$

(oxidation no. of Cl) + 3(oxidation no. of O) = -1
(oxidation no. of Cl) + 3(-2) = -1
(oxidation no. of Cl) = +5

d. $ClO_4^-$

(oxidation no. of Cl) + 4(oxidation no. of O) = -1
(oxidation no. of Cl) + 4(-2) = -1
(oxidation no. of Cl) = +7

e. HCl

(oxidation no. of H) + (oxidation no. of Cl) = 0
(+1) + (oxidation no. of Cl) = 0
(oxidation no. of Cl) = -1

f. $Cl_2O$

2(oxidation no. of Cl) + oxidation no. of O = 0
2(oxidation no. of Cl) + (-2) = 0
2(oxidation no. of Cl) = +2
(oxidation no. of Cl) = +1

Practice Problem B

| | | | |
|---|---|---|---|
| a. | $Zn(OH)_2$ | ionic | zinc hydroxide |
| b. | $CuI_2$ | ionic | copper(II) iodide |
| c. | $P_2O_5$ | molecular | diphosphorus pentoxide |
| d. | $HNO_2$ | molecular | nitrous acid |
| e. | $Se_2Br_2$ | molecular | diselenium dibromide |
| f. | $Pb(CO_3)_2$ | ionic | lead(IV) carbonate |
| g. | $Li_3N$ | ionic | lithium nitride |
| h. | $H_3P$ | molecular | hydrophosphoric acid |
| i. | $CaCrO_4$ | ionic | calcium chromate |
| j. | $AsCl_5$ | molecular | arsenic pentachloride |

Practice Problem C

| | | |
|---|---|---|
| a. | xenon tetrafluoride | $XeF_4$ |
| b. | magnesium chlorate | $Mg(ClO_3)_2$ |
| c. | hydroselenic acid | $H_2Se$ |
| d. | nickel(II) sulfide | $NiS$ |
| e. | dinitrogen trioxide | $N_2O_3$ |
| f. | hypobromous acid | $HBrO$ |
| g. | copper(I) sulfate | $Cu_2SO_4$ |

|    |                  |            |
|----|------------------|------------|
| h. | ammonium nitrate | $NH_4NO_3$ |
| i. | phosphorous acid | $H_3PO_3$  |
| j. | aluminum oxide   | $Al_2O_3$  |

## ANSWERS TO SELF-TEST

1. a
2. c
3. b
4. c
5. d
6. c
7. b
8. d
9. c
10. d

# CALCULATIONS BASED ON CHEMICAL FORMULAS

## CHAPTER 8

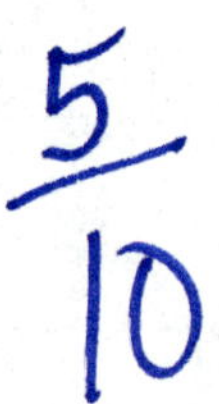

---

### SUMMARY OUTLINE

**8.1**   <u>The Chemical Mole</u>

A mole is defined as $6.02 \times 10^{23}$ objects or particles.

$6.02 \times 10^{23}$ is known as Avogadro's number.

**8.2**   <u>Molar Mass</u>

One mole of a substance has a mass corresponding to the formula weight expressed in grams.  This mass is called the molar mass.

Molar mass has units of $\dfrac{g}{mol}$.

One mole of any element has the same number of atoms ($6.02 \times 10^{23}$) as one mole of any other element.  No two elements have the same molar mass.

One mole of molecules contains $6.02 \times 10^{23}$ molecules and a multiple of $6.02 \times 10^{23}$ atoms corresponding to the subscripts in the formula.

Formula weights of molecular compounds are called molecular weights.

**8.3**   <u>Calculation of Molar Masses</u>

To calculate the molar mass of a substance, find the formula weight and express the units in grams.

**8.4   Additional Calculations Based on the Mole Concept**

1 mole of substance = formula weight in grams
1 mole of substance = $6.02 \times 10^{23}$ particles.

Using formula weights and Avogadro's number, masses and number of particles can be interconverted.

$$\text{g} \xleftrightarrow{\text{(molar mass)}} \text{mole} \xleftrightarrow{(6.02 \times 10^{23})} \text{particles}$$

**8.5   Empirical and True Formulas**

The empirical formula of a compound is the simplest whole-number ratio of ions or atoms in a compound.

The molar mass calculated from the empirical formula is called the apparent molar mass.

The true formula represents the total number of ions or atoms present in a formula unit of the compound.

**8.6   Calculation of the Empirical Formula**

To calculate the empirical formula,

1.   Assume a 100g sample and convert the percentage of each element to actual mass.
2.   Convert the mass of each element to moles.
3.   Find the smallest whole-number ratio of the elements.

**8.7   Calculation of the True Formula from the Empirical Formula**

The true formula can be calculated using the following relationship,

$$n = \frac{\text{actual molar mass}}{\text{apparent molar mass}}$$

where n represents a whole-number multiple of the empirical formula.  Multiply the subscripts in the empirical formula by n to find the true formula.

To calculate the true formula of the compound when actual molar mass and percentages (or masses) of the elements are given,

1.   Calculate the empirical formula.
2.   Calculate the apparent molar mass.
3.   Find n, the whole-number multiple of the empirical formula.
4.   Multiply the subscripts in the empirical formula by n.

---

SOLUTIONS TO SELECTED TEXT STUDY QUESTIONS AND PROBLEMS
and
PRACTICE PROBLEMS

---

7.   Complete the following table.

| | Element | Weight(g) | Number of Moles | Number of Atoms |
|---|---|---|---|---|
| a. | Rb | 14.9 | | |
| b. | Kr | | | $1.29 \times 10^{20}$ |
| c. | N | | 1.23 | |
| d. | Cr | | | $4.08 \times 10^{39}$ |
| e. | S | 40.5 | | |
| f. | Mo | | 7.86 | |

Use the following "mole map" to interconvert weights, number of moles and number of atoms:

$$\text{g} \quad \underset{\text{(molar mass)}}{\longleftrightarrow} \quad \text{mol} \quad \underset{(6.02 \times 10^{23})}{\longleftrightarrow} \quad \text{particles}$$

Multiply the given information by the appropriate unit-factors to find the required information.

a.   Rb

14.9g Rb $\longrightarrow$ ? mole Rb

To convert from g $\longrightarrow$ mol, the mole map shows that only one conversion factor is required.  Arrange the units in the conversion factor to cancel g Rb and introduce mol Rb in the numerator.

$$14.9\text{g Rb} \times \frac{?\ \text{mol Rb}}{?\ \ \text{g Rb}}$$

Find the molar mass of Rb from the atomic weight of Rb found on the periodic table.  This mass expressed in grams is the weight of 1 mol Rb.

$$14.9\text{g Rb} \times \frac{1\ \text{mol Rb}}{85.5\text{g Rb}} = 0.174269\ \text{mol Rb}$$
$$= 0.174\ \text{mol Rb}$$

Be sure to include both units and labels (in this case, Rb) with all numbers.  The answer is limited to 3 significant figures from the given mass.

14.9g Rb $\longrightarrow$ ? atoms Rb

To convert from g $\longrightarrow$ atoms, the "mole map" shows a 2 step conversion, g $\longrightarrow$ mol $\longrightarrow$ atoms.

$$14.9\text{g Rb} \times \frac{?\ \text{mol Rb}}{?\ \ \text{g Rb}} \times \frac{?\ \text{atoms Rb}}{?\ \text{mol Rb}}$$

$$14.9\text{g Rb} \times \frac{1\ \text{mol Rb}}{85.5\text{g Rb}} \times \frac{6.02 \times 10^{23}\ \text{atoms Rb}}{1\ \text{mol Rb}} = 1.05 \times 10^{23}\ \text{atoms Rb}$$

Consider the mathematical solution to this problem.  After cancelation of the units, the problem reads

$$\frac{(14.9)\ \times\ (6.02 \times 10^{23})}{85.5}\ \text{atoms Rb} =$$

A scientific calculator can be used for calculations involving numbers in scientific notation.  The correct key

to use will be marked EE , Exp , or EEX .  To enter the
number $6.02 \times 10^{23}$, first enter 6.02, press the EE key, and
then press 23.  The calculator will display this as
6.02 23.

To enter the numbers in the problem above, press 14.9 X
6.02 EE 23 ÷ 85.5 = .  The display reads 1.049099 23.
The answer must be rounded off to 3 significant figures
and reported as $1.05 \times 10^{23}$ atoms Rb.

b.  Kr

$1.29 \times 10^{20}$ atoms Kr $\longrightarrow$ ? mol Kr

$$1.29 \times 10^{20} \text{ atoms Kr} \times \frac{\text{? mol Kr}}{\text{? atoms Kr}} =$$

$$1.29 \times 10^{20} \text{ atoms Kr} \times \frac{1 \text{ mol Kr}}{6.02 \times 10^{23} \text{ atoms Kr}} = 2.14 \times 10^{-4} \text{ mol Kr}$$

To solve the problem using a calculator, press 1.29 EE
20 ÷ 6.02 EE 23 = .  The display reads 2.142857 -04.
The answer must be rounded off to 3 significant figures
and reported as $2.14 \times 10^{-4}$ mol Kr.

$1.29 \times 10^{20}$ atoms Kr $\longrightarrow$ ? g Kr

Beginning with atoms Kr, the "mole map" shows that a
two-step conversion is required,

$$\text{atoms} \longrightarrow \text{mol} \longrightarrow \text{g}$$

$$1.29 \times 10^{20} \text{ atoms Kr} \times \frac{\text{? mol Kr}}{\text{? atoms Kr}} \times \frac{\text{? g Kr}}{\text{? mol Kr}}$$

$$1.29 \times 10^{20} \text{ atoms Kr} \times \frac{1 \text{ mol Kr}}{6.02 \times 10^{23} \text{ atoms Kr}} \times \frac{83.8 \text{g Kr}}{1 \text{ mol Kr}} = 1.7957 \times 10^{-2} \text{g Kr}$$

$$= 1.80 \times 10^{-2} \text{g Kr}$$

To solve the problem using a calculator, press 1.29 EE
20 ÷ 6.02 EE 23 X 83.8 = .  The display reads
1.795714 - 02.  The answer must be rounded to 3 significant
figures and reported as $1.80 \times 10^{-2}$g Kr.

c.  N

1.23 mol N $\longrightarrow$ ? g

$$1.23 \text{ mol N} \times \frac{\text{? g N}}{\text{? mol N}}$$

$$1.23 \text{ mol N} \times \frac{14.0 \text{g N}}{1 \text{ mol N}} = 17.2 \text{g N}$$

$$1.23 \text{ mol N} \longrightarrow \text{? atoms N}$$

$$1.23 \text{ mol N} \times \frac{\text{? atom N}}{\text{? mol N}}$$

$$1.23 \text{ mol N} \times \frac{6.02 \times 10^{23} \text{ atoms N}}{1 \text{ mol N}} = 7.4046 \times 10^{23} \text{ atoms N}$$
$$= 7.40 \times 10^{23} \text{ atoms N}$$

Consult Appendix C in the text for answers to parts d, e, and f.

***Practice Problem A

Complete the following calculations in the space provided.

a.  How many moles in 10.0g of Ca?

b.  How many grams do $3.75 \times 10^{30}$ atoms of Ca weigh?

c.  How many atoms in $5.35 \times 10^{-1}$ mol of Ca?

9.    For each of the following compounds, give the number of cations and the number of anions present in 3.50 moles of the compound

a.  $Na_2Cr_2O_7$            b.  $HgC_2O_4$            c.  $K_2SO_4$

To find moles of a particular element (or ion) within a compound, utilize the formula subscript for that particular element (or ion).

In part a above, 1 mol $Na_2Cr_2O_7$ contains 2 mol $Na^+$ and 1 mol $Cr_2O_7^{2-}$.  These relationships, found from the subscripts in the formula, can be used as unit-factors.

Calculating molar quantities for elements or ions found within compounds is often referred to as calculating "moles within moles".  The mole map can be expanded to compare component elements to their compounds.  (The subscript $\underline{E}$ refers to element and the subscript $\underline{C}$ refers to compound).

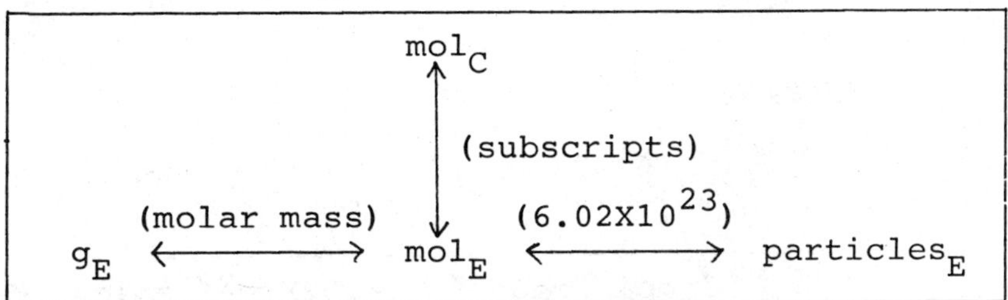


Sometimes masses of compounds are given and masses of component elements are sought.  The following map includes these relationships.

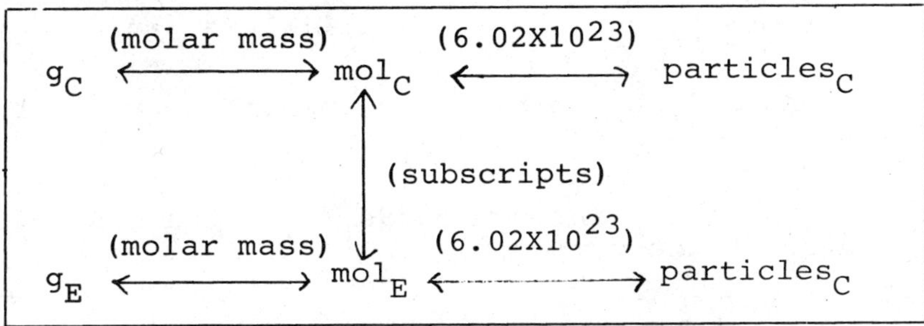


The key to solving these problems will be the mole relationship between the element and the compound.

a.   3.50 mol $Na_2Cr_2O_7 \longrightarrow$ ? cations $Na^+$

The mole map can be used in the following way,



$$3.50 \text{ mol } Na_2Cr_2O_7 \times \frac{? \text{ mol } Na^+}{? \text{ mol } Na_2Cr_2O_7} \times \frac{? \text{ cations } Na^+}{? \text{ mol } Na^+} =$$

$$3.50 \text{ mol } Na_2Cr_2O_7 \times \frac{2 \text{ mol } Na^+}{1 \text{ mol } Na_2Cr_2O_7} \times \frac{6.02 \times 10^{23} \text{ cations } Na^+}{1 \text{ mol } Na^+} =$$

$$4.214 \times 10^{24} \text{ cations } Na^+ =$$

$$4.21 \times 10^{24} \text{ cations } Na^+ =$$

$$3.50 \text{ mol } Na_2Cr_2O_7 \longrightarrow ? \text{ anions } Cr_2O_7{}^{2-}$$

The "mole map" indicates a two-step conversion,

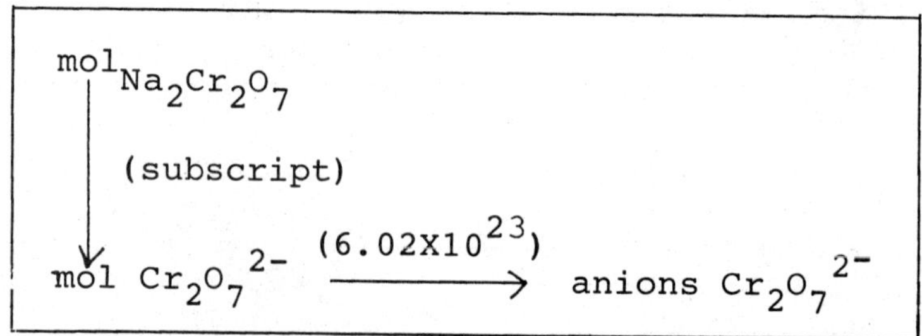

$$3.50 \text{ mol } Na_2Cr_2O_7 \times \frac{1 \text{ mol } Cr_2O_7{}^{2-}}{1 \text{ mol } Na_2Cr_2O_7} \times \frac{6.02 \times 10^{23} \text{ anions } Cr_2O_7{}^{2-}}{1 \text{ mol } Cr_2O_7{}^{2-}} =$$

(Notice that there are twice as many cations of $Na^+$ as anions of $Cr_2O_7{}^{2-}$.)

$$2.107 \times 10^{24} \text{ anions } Cr_2O_7{}^{2-} =$$

$$2.11 \times 10^{24} \text{ anions } Cr_2O_7{}^{2-}$$

Consult Appendix C in the text for answers to parts b and c.

***Practice Problem B

For each of the following compounds, give the mass of sulfur in 3.50 moles of the compound.

a.  $H_2SO_4$

$$3.50 \times \frac{1 \text{ moles}}{1 \text{ mole } H_2SO_4} \times \frac{32.1}{1 \text{ mol } S} = 112 \text{ g S}$$

b.  $Al_2S_3$

$$3.50 \times \frac{3 \text{ moles}}{1 \text{ mole } Al_2S_3} \times \frac{32.1}{1 \text{ mol } S} = 337 \text{ g S}$$

c.    $PbS_2$

$$3.50 \times \frac{2 \text{ mol}}{1 \text{ mol}} \times \frac{32.1}{1 \cdot \text{mol}} = 225g$$

*labels?* ✓

15.    Sodium hypochlorite, NaClO, is the active ingredient in house-hold bleach.  What is the mass of $6.57 \times 10^{17}$ formula units of NaClO?

Use the "mole map" as a guide to interconverting weight, number of moles and number of molecules or formula units.

$$\text{g} \xleftrightarrow{\text{(molar mass)}} \text{mol} \xleftrightarrow{(6.02 \times 10^{23})} \text{particles}$$

For this problem, formula units are given and g are required. The "mole map" indicates a two-step conversion.

$$\text{formula units} \xrightarrow{(6.02 \times 10^{23})} \text{mol} \xrightarrow{\text{(molar mass)}} \text{g}$$

The molar mass must be calculated by adding up the formula weight of the compound.  The formula weight expressed in grams for any compound is equal to one mole of that compound.

$$\text{FW NaClO} = (1 \text{ atom Na})\left(\frac{23.0 \text{ amu}}{1 \text{ atom Na}}\right) + (1 \text{ atom Cl})\left(\frac{35.5 \text{ amu}}{1 \text{ atom Cl}}\right)$$

$$+ (1 \text{ atom O})\left(\frac{16.0 \text{ amu}}{1 \text{ atom O}}\right)$$

$$= 23.0 \text{ amu} + 35.5 \text{ amu} + 16.0 \text{ amu}$$

$$= 74.5 \text{ amu}$$

The molar mass of NaClO is 74.5 g.

$$6.57 \times 10^{17} \text{ formula units NaClO} \times \frac{1 \text{ mol NaClO}}{6.02 \times 10^{23} \text{ formula units NaClO}} \times \frac{74.5 \text{g NaClO}}{1 \text{ mol NaClO}} =$$

$$8.13 \times 10^{-5} \text{g NaClO}$$

To solve the problem using a calculator, press 6.57 $\boxed{\text{EE}}$ 17 $\boxed{\div}$ 6.02 $\boxed{\text{EE}}$ 23 $\boxed{\text{X}}$ 74.5 $\boxed{=}$ .  The display reads 8.130648 -05.  The answer must be rounded to 3 significant figures and reported as $8.13 \times 10^{-5}$g NaClO.

***Practice Problem C

a.  How many molecules of $SOCl_2$ are present in a 15.5g sample of $SOCl_2$?

b.  How many grams of sulfur are present in a 15.5g sample of $SOCl_2$?

27.  Hydrazine, a liquid used as a rocket fuel, consists of 87.5% nitrogen and 12.5% hydrogen and has a molar mass of 32.0g/mol. What is the true formula of hydrazine?

Follow the guidelines given in section 8.6 for finding true formulas.

First calculate the empirical formula.

Assume 100g sample.
wt N = (0.875 X 100g) = 87.5g N
wt H = (0.125 X 100g) = 12.5g H

$$\text{mole N} = 87.5g\ N \times \frac{1\ mol\ N}{14.0g\ N} = 6.25\ mol\ N$$

$$\text{mole H} = 12.5g\ H \times \frac{1\ mol\ H}{1.01g\ H} = 12.4\ mol\ O$$

$$N: \frac{6.25}{6.25} = 1.00$$

$$H: \frac{12.4}{6.25} = 1.98$$

Therefore, the empirical formula is $NH_2$

Find the apparent molar mass.

Apparent FW = (1 atom N) $\left(\dfrac{14.0 \text{ amu}}{1 \text{ atom N}}\right)$ + (2 atom H) $\left(\dfrac{1.01 \text{ amu}}{1 \text{ atom H}}\right)$

$$= 14.0 \text{ amu} + 2.02 \text{ amu}$$

$$= 16.0 \text{ amu}$$

The apparent molar mass is 16.0 g/mol

Find n, the whole-number multiple of the empirical formula.

$$n = \frac{\text{actual molar mass}}{\text{apparent molar mass}}$$

$$= \frac{32.0 \text{ g/mol}}{16.0 \text{ g/mol}} = 2.00$$

Multiply the subscripts in the empirical formula by n.

$$\text{true formula} = N_{(1\times2)}H_{(2\times2)} = N_2H_4$$

***Practice Problem D

Aspirin contains 60.0% carbon, 4.48% hydrogen, and 35.5% oxygen. The molar mass of aspirin is 180 g/mol.  What is the true formula of aspirin?

---



---

Circle the correct answer in the following multiple choice questions.

1.  How many grams are contained in 1.00 mol of baking soda, $NaHCO_3$ (sodium hydrogen carbonate)?

    a.  52.0 g

    b.  108.0 g

    c.  84.0 g

    d.  1.00 g

2.     The average mass of one carbon atom is
       a.  $7.22X10^{24}$ g
       b.  $2.00X10^{-23}$ g
       c.  12.0 g
       d.  $5.0X10^{22}$ g

3.     The number of hydrogen atoms in 10.0 g of glucose, $C_6H_{12}O_6$ is
       a.  $4.01X10^{23}$
       b.  $3.34X10^{22}$
       c.  $2.78X10^{21}$
       d.  $1.30X10^{31}$

4.     The mass of $10^6$ aspirin molecules, $C_9H_8O_4$ is
       a.  $1.66X10^{-18}$ g
       b.  $2.99X10^{-15}$ g
       c.  $2.99X10^{-16}$ g
       d.  $1.08X10^{32}$ g

5.     The number of moles of nitrogen atoms in 100.0 g of nitro-
       glycerin, $C_3H_5(NO_3)_3$ is
       a.  0.1468 moles
       b.  0.4405 moles
       c.  2.913 moles
       d.  1.322 moles

6.     Which of the following contains the most molecules?
       a.  10.0 g $H_2O$
       b.  10.0 g $CO_2$
       c.  10.0 g Na
       d.  10.0 g $F_2$

7.     A sample of 2.2 g of $CO_2$ contains
       a.  0.05 mol $CO_2$, 0.10 mol C, 0.05 mol O
       b.  0.05 mol $CO_2$, 0.05 mol C, 0.10 mol O
       c.  1.00 mol $CO_2$, 1.00 mol C, 2.00 mol O
       d.  20 mol $CO_2$, 20  mol C, 10  mol O

8.     A compound of S and P, used on match tips, consists of 43.7%
       S and 56.3% P.  The empirical formula is
       a.  $P_3S_4$
       b.  $P_3S_2$
       c.  $P_2S$
       d.  $P_4S_3$

9.   The empirical formula of a compound is $C_2HNO_2$ and the molar
     mass is 210 g/mol.  The true formula is

     a.   $C_2HNO_2$

     b.   $C_4H_2N_2O_4$

     c.   $C_6H_3N_3O_6$

     d.   $C_3HNO_3$

10.  Acetone, used as a mail polish remover, consists of 62.0% C,
     10.4% H, and 27.5% O by mass.  The molar mass is 58.1 g/mol.
     The true formula is

     a.   $C_2H_2O_2$

     b.   $C_3H_6O$

     c.   $C_2H_{18}O$

     d.   $C_2HO_3$

---

## ANSWERS TO PRACTICE PROBLEMS

Practice Problem A

a.
$$g\ Ca \xrightarrow{\text{(molar mass)}} mol\ Ca$$

$$10.0g\ Ca \times \frac{1\ mol\ Ca}{40.1\ g\ Ca} = .249\ mol\ Ca$$

b.
$$atoms\ Ca \xrightarrow{(6.02\times10^{23})} mol\ Ca \xrightarrow{\text{(molar mass)}} g\ Ca$$

$$3.75\times10^{30}\ atoms\ Ca \times \frac{1\ mol\ Ca}{6.02\times10^{23}\ atoms\ Ca} \times \frac{40.1\ g\ Ca}{1\ mol\ Ca} = 2.50\times10^{8}\ g\ Ca$$

c.
$$mol\ Ca \xrightarrow{(6.02\times10^{23})} atoms\ Ca$$

$$5.35\times10^{-1}\ mol\ Ca \times \frac{6.02\times10^{23}\ atoms\ Ca}{1\ mol\ Ca} = 3.22\times10^{23}\ atoms\ Ca$$

Practice Problem B

a.   Refer to the "mole map" on page

$$mol\ H_2SO_4$$
$$\downarrow$$
$$mol\ S \longrightarrow g\ S$$

$$3.50\ mol\ H_2SO_4 \times \frac{1\ mol\ S}{1\ mol\ H_2SO_4} \times \frac{32.1\ g\ S}{1\ mol\ S} = 112\ g\ S$$

b.

$$mol\ Al_2S_3$$
$$\downarrow$$
$$mol\ S \longrightarrow g\ S$$

$$3.50\ \cancel{mol\ Al_2S_3} \times \frac{3\ \cancel{mol\ S}}{1\ \cancel{mol\ Al_2S_3}} \times \frac{32.1\ g\ S}{1\ \cancel{mol\ S}} = 337\ g\ S$$

c.

$$mol\ PbS_2$$
$$\downarrow$$
$$mol\ S \longrightarrow g\ S$$

$$3.50\ mol\ PbS_2 \times \frac{2\ mol\ S}{1\ \cancel{mol\ PbS_2}} \times \frac{32.1\ g\ S}{1\ mol\ S} = 225\ g\ S$$

**Practice Problem C**

a.  $FW\ SOCl_2 = (1\ atom\ S)\left(\dfrac{32.1\ amu}{1\ atom\ S}\right) + (1\ atom\ O)\left(\dfrac{16.0\ amu}{1\ atom\ O}\right)$

$$+\ (2\ atom\ Cl)\left(\frac{35.5\ amu}{1\ atom\ Cl}\right)$$

$$=\ 32.1\ amu + 16.0\ amu + 71.0\ amu$$

$$=\ 119.1\ amu$$

The molar mass of $SOCl_2$ is 119.1 g/mol.

$$g \longrightarrow mol \longrightarrow molecules$$

$$15.5\ g\ \cancel{SOCl_2} \times \frac{1\ \cancel{mol\ SOCl_2}}{119.1\ g\ \cancel{SOCl_2}} \times \frac{6.02\times10^{23}\ molecules\ SOCl_2}{1\ \cancel{mol\ SOCl_2}} = 7.83\times10^{22}$$
$$molecules\ SOCl_2$$

b.  Refer to the "mole map" in the middle of page 109.

$$g_{SOCl_2} \longrightarrow mol_{SOCl_2}$$
$$\downarrow$$
$$g_S \longleftarrow mol_S$$

$$15.5g\ SOCl_2 \times \frac{1\ mol\ SOCl_2}{119.1g\ SOCl_2} \times \frac{1\ mol\ S}{1\ mol\ SOCl_2} \times \frac{32.1g\ S}{1\ mol\ S} = 4.18g\ S$$

**Practice Problem D**

Assume 100 g sample
wt C = (.600 X 100 g)  = 60.0 g C
wt H = (.0448 X 100 g) =  4.48 g H
wt O = (.355 X 100 g)  = 35.5 g O

$$mol\ C = 60.0\ g\ C \times \frac{1\ mol\ C}{12.0\ g\ C} = 5.00\ mol\ C$$

$$\text{mol H} = 4.48 \text{ g H} \times \frac{1 \text{ mol H}}{1.01 \text{ g H}} = 4.44 \text{ mol H}$$

$$\text{mol O} = 35.5 \text{ g O} \times \frac{1 \text{ mol O}}{16.0 \text{ g O}} = 2.22 \text{ mol O}$$

$$\text{C} : \frac{5.00}{2.22} = 2.25$$

$$\text{H} : \frac{4.44}{2.22} = 2.00$$

$$\text{O} : \frac{2.22}{2.22} = 1.00$$

Multiply the ratio by 4 to achieve a whole-number ratio.

$$C = 2.25 \times 4 = 9.00$$
$$H = 2.00 \times 4 = 8.00$$
$$O = 1.00 \times 4 = 4.00$$

The empirical formula is $C_9H_8O_4$.

The apparent FW = (9 atom C) $\left(\dfrac{12.0 \text{ amu}}{1 \text{ atom C}}\right)$ +

(8 atom H) $\left(\dfrac{1.01 \text{ amu}}{1 \text{ atom H}}\right)$ +

(4 atom O) $\left(\dfrac{16.0 \text{ amu}}{1 \text{ atom O}}\right)$

$$= \quad 108 \text{ amu} + 8.08 \text{ amu} + 64.0 \text{ amu}$$

$$= \quad 180 \text{ amu}$$

The apparent molar mass is 180 g/mol.

The molar mass is 180 g/mol so the true formula is also $C_9H_8O_4$.

---

## ANSWERS TO SELF-TEST

1.  c
2.  b
3.  a
4.  c
5.  d
6.  a
7.  b
8.  d
9.  c
10.  b

# CHEMICAL EQUATIONS

# CHAPTER 9

---

SUMMARY OUTLINE

---

### 9.1 The Language of Chemical Equations

A chemical equation uses symbols of elements, and formulas of compounds to describe a chemical reaction.

$$\text{one or more reactants} \xrightarrow{\text{(yield)}} \text{one or more products}$$

The numbers in front of the formulas are coefficients that tell how many formula units of a substance are present.

### 9.2 Conservation of Mass in Chemical Reactions

All chemical reactions follow the law of conservation of mass. The total mass of reactants is equal to the total mass of products (Figure 9-1).

### 9.3 Balancing Chemical Equations

A balanced chemical equation has the same number of each kind of atom on each side of the arrow.

The following rules can be used to balance chemical equations:

1. Write correct formulas for reactants and products.
2. Count the number of atoms of each element in the reactants and products.
3. Using smallest possible whole numbers, insert co-efficients for one formula at a time to balance the number of atoms of each element in the reactants and products.

4. Recount atoms on each side of the equation.
5. If any element remains "unbalanced", change coefficients until the equation is balanced.

## 9.4   Types of Chemical Reactions

In a decomposition reaction, a single reactant is converted
into two or more simpler products.

$$AB \longrightarrow A + B$$

In a combination reaction, two or more reactants are combined
to form one product.

$$A + B \longrightarrow AB$$

In a single replacement reaction, an ion, atom, or group of
atoms replaces another ion, atom, or group of atoms in a
compound.

$$A + BC \longrightarrow AC + B$$

In a double replacement reaction, two compounds exchange parts
to form two new compounds.

$$AB + CD \longrightarrow AD + CB$$

## 9.5   Stoichiometry

The numerical coefficients of a balanced chemical equation
also tell the number of moles of reactants and the number of
moles of products.

Stoichiometry calculations use molar relationships in a
balanced chemical equation to compare quantities of reactants
and products.

A mole ratio is the ratio of moles of any two substances
determined from the numerical coefficients of a balanced
equation.

## 9.6   Stoichiometric Calculations

There are four categories of stoichiometric calculations.  In
each category, a given quantity of substance A is used to find
an unknown quantity of substance B.

A mole-mole stoichiometry problem involves,

$$X \text{ mol } A \longrightarrow ? \text{ mol } B.$$

A mole-weight stoichiometry problem involves,

$$X \text{ mol } A \longrightarrow ? \text{ g } B.$$

A weight-mole stoichiometry problem involves,

$$X \text{ g } A \longrightarrow ? \text{ mol } B.$$

A weight-weight stoichiometry problem involves,

$$X \text{ g } A \longrightarrow ? \text{ g } B.$$

The fundamental relationships important to stoichiometric calculations are summarized in the following "mole map":

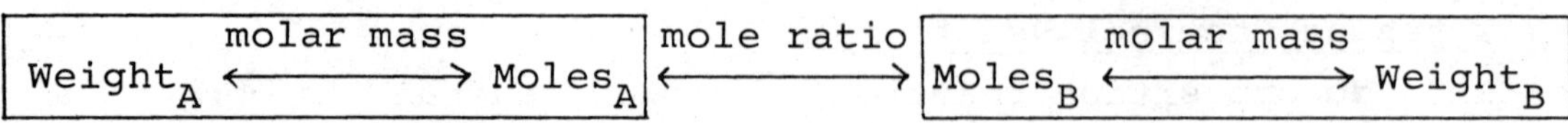


## 9.7    Percentage Yield

The theoretical yield is the maximum amount of product that can be produced based on stoichiometric calculations.

The actual yield is the amount of product actually recovered at the end of an experiment.  The actual yield is seldom equal to the theoretical yield.

Percentage yield compares the actual yield to the theoretical yield.

$$\text{\% yield} = \frac{\text{actual yield}}{\text{theoretical yield}} \times 100\%$$

## 9.8    Limiting Reactant

The limiting reactant is the substance that is completely consumed by a reaction with an excess of other reactants.  The limiting reactant limits the amount of product that can form.

---

SOLUTIONS TO SELECTED STUDY QUESTIONS AND PROBLEMS
and
PRACTICE PROBLEMS

---

16.    For the following equation,

$$3KOH_{(aq)} + H_3PO_{4(aq)} \longrightarrow K_3PO_{4(aq)} + 3H_2O_{(l)}$$

a.    How many moles of $H_3PO_4$ are required to react with 2.15 moles of KOH?

This is a mole-mole stoichiometry problem.  The mole map in section 9.6 shows that only one conversion is required.

g KOH $\xrightarrow{\quad\times\quad}$ 2.15 mol KOH $\longrightarrow$ ? mol $H_3PO_4$ $\xrightarrow{\quad\times\quad}$ g $H_3PO_4$

Step 1:  Write down the given number of moles.

2.15 mol KOH

Step 2:  Multiply the given number of moles by the mole ratio to find moles of the required substance.

$$(2.15 \text{ mol KOH})\left(\frac{\text{? mol } H_3PO_4}{\text{? mol KOH}}\right) =$$

This arrangement of the units in the unit-factor allows cancellation of mol KOH.  Be sure to include the labels, KOH and $H_3PO_4$, with the units mole.

The mole ratio needed for this particular unit-factor can be found from the coefficients in the equation.

$$(2.15 \text{ mol KOH})\left(\frac{1 \text{ mol } H_3PO_4}{3 \text{ mol KOH}}\right) = 0.716667 \text{ mol } H_3PO_4$$
$$= 0.717 \text{ mol } H_3PO_4$$

Notice that the numbers in mole ratios are exact and do not limit the number of significant figures.

b.   How many grams of $K_3PO_4$ will be produced by 17.8 g of $H_3PO_4$?

This is a weight-weight stoichiometry problem.   The mole map,

$$17.8g\ H_3PO_4 \longrightarrow ?\ \text{mol } H_3PO_4 \longrightarrow ?\ \text{mol } K_3PO_4 \longrightarrow ?\ g\ K_3PO_4$$

shows that three conversions are required.   The four steps outlined below accomplish these conversions in individual steps.

Step 1:   Write down the given weight.

$$17.8g\ H_3PO_4$$

Step 2:   Convert the given weight to moles.

$$(17.8g\ H_3PO_4)\left(\frac{?\ \text{mol } H_3PO_4}{?\ g\ H_3PO_4}\right) =$$

Add up the formula weight of $H_3PO_4$.   The formula weight expressed in grams is the mass of one mole of the compound.

$$(17.8g\ H_3PO_4)\ \frac{1 \text{ mol } H_3PO_4}{98.0g\ H_3PO_4} = 0.182 \text{ mol } H_3PO_4$$

Step 3:   Convert moles of $H_3PO_4$ to moles of $K_3PO_4$.

$$0.182 \text{ mol } H_3PO_4 \times \frac{?\ \text{mol } K_3PO_4}{?\ \text{mol } H_3PO_4} =$$

The balanced chemical equation shows that 1 mol $H_3PO_4$ can produce 1 mol $K_3PO_4$.

$$0.182 \text{ mol } H_3PO_4 \times \frac{1 \text{ mol } K_3PO_4}{1 \text{ mol } H_3PO_4} = 0.182 \text{ mol } K_3PO_4$$

Step 4:   Convert moles of $K_3PO_4$ to weight of $K_3PO_4$.

$$0.182 \text{ mol } K_3PO_4 \times \frac{?\ g\ K_3PO_4}{?\ \text{mol } K_3PO_4} =$$

Find the formula weight of $K_3PO_4$.  The formula weight expressed in grams is the molar mass.

$$0.182 \text{ mol } K_3PO_4 \times \frac{212.3g \ K_3PO_4}{1 \text{ mol } K_3PO_4} = 38.6386g \ K_3PO_4 = 38.6g \ K_3PO_4$$

The answer is limited to 3 significant figures from the given mass of $H_3PO_4$.  These four steps can be combined as follows:

$$(17.8g \ H_3PO_4)\left(\frac{1 \text{ mol } H_3PO_4}{98.0g \ H_3PO_4}\right)\left(\frac{1 \text{ mol } K_3PO_4}{1 \text{ mol } H_3PO_4}\right)\left(\frac{212.3g \ K_3PO_4}{1 \text{ mol } K_3PO_4}\right) = 38.6g \ K_3PO_4$$

c.  How many moles of KOH will be required to react with 24.6g of $H_3PO_4$?

This is a weight-mole stoichiometry problem.  The mole map,

$$24.6g \ H_3PO_4 \longrightarrow ? \text{ mol } K_3PO_4 \longrightarrow ? \text{ mol KOH} \xrightarrow{\quad\times\quad} g \text{ KOH}$$

shows that two conversions are required.  The three steps outlined below accomplish these conversions in individual steps.

Step 1:  Write down the given weight.

$$24.6g \ H_3PO_4$$

Step 2:  Convert the given weight to moles.

$$24.6g \ H_3PO_4 \times \frac{1 \text{ mol } H_3PO_4}{98.0g \ H_3PO_4} = 0.251 \text{ mol } H_3PO_4$$

Step 3:  Convert moles of $H_3PO_4$ to moles of KOH.

$$0.251 \text{ mol } H_3PO_4 \times \frac{3 \text{ mol KOH}}{1 \text{ mol } H_3PO_4} = 0.753 \text{ mol KOH}$$

These three steps can be combined as follows:

$$24.6g \ H_3PO_4 \times \frac{1 \text{ mol } H_3PO_4}{98.0g \ H_3PO_4} \times \frac{3 \text{ mol KOH}}{1 \text{ mol } H_3PO_4} = 0.753 \text{ mol KOH}$$

d.  How many grams of $H_2O$ will be produced when 34.7g of $K_3PO_4$ are formed?

This is a weight-weight stoichiometry problem.  The mole map,

$$34.7g \ K_3PO_4 \longrightarrow ? \text{ mol } K_3PO_4 \longrightarrow ? \text{ mol } H_2O \longrightarrow ? \ g \ H_2O$$

shows that three conversions are required.  The four steps outlined below accomplish these conversions in individual steps.

Step 1:  Write down the given weight.

$$34.7g \ K_3PO_4$$

Step 2:   Convert the given weight to moles.

$$34.7g\ K_3PO_4 \times \frac{1\ mol\ K_3PO_4}{212.3g\ K_3PO_4} = 0.163\ mol\ K_3PO_4$$

Step 3:   Convert moles of $K_3PO_4$ to moles of $H_2O$.

$$0.163\ mol\ K_3PO_4 \times \frac{3\ mol\ H_2O}{1\ mol\ K_3PO_4} = 0.489\ mol\ H_2O$$

Step 4:   Convert moles of $H_2O$ to weight of $H_2O$.

$$0.489\ mol\ H_2O \times \frac{18.0g\ H_2O}{1\ mol\ H_2O} = 8.80g\ H_2O$$

These four steps can be combined as follows:

$$34.7g\ K_3PO_4 \times \frac{1\ mol\ K_3PO_4}{212.3g\ K_3PO_4} \times \frac{3\ mol\ H_2O}{1\ mol\ K_3PO_4} \times \frac{18.0g\ H_2O}{1\ mol\ H_2O} = 8.83g\ H_2O$$

The answers differ in the two methods, due to the rounding off of intermediate answers in the four step method.

25.   How many grams of aluminum are needed to react with sulfur to form 600.0g of aluminum sulfide, $Al_2S_{3(S)}$ ?

Write a balanced chemical equation using correct formulas for reactants and products.

Step 1:   Write the initial equation.

$$Al_{(S)} + S_{(S)} \longrightarrow Al_2S_{3(S)}$$

Step 2:   Count the atoms of each element.

| element | reactants | products |
|---------|-----------|----------|
| Al | 1 | 2 |
| S | 1 | 3 |

Step 3:   Insert coefficients for aluminum

$$2Al_{(S)} + S_{(S)} \longrightarrow Al_2S_{3(S)}$$

Insert coefficients for sulfur.

$$2Al_{(S)} + 3S_{(S)} \longrightarrow Al_2S_{3(S)}$$

Step 4:   Recount atoms.

| element | reactants | products |
|---------|-----------|----------|
| Al | 2 | 2 |
| S | 3 | 3 |

Step 5:  No change in coefficients is necessary.

Use the balanced chemical equation to find grams of aluminum needed to product 600.0g of aluminum sulfide.  An effective bookkeeping tool for solving stoichiometry problems is to label the chemical equation with the given and required quantities.

$$2Al_{(S)} + S_{(S)} \longrightarrow Al_2S_{3\,(aq)}$$

$$?g \qquad\qquad\qquad 600.0g$$

From the labels, it is clear that this is a weight-weight stoichiometry problem.

The mole map,

$$600.0g\ Al_2S_3 \longrightarrow ?\ mol\ Al_2S_3 \longrightarrow ?\ mol\ Al \longrightarrow ?\ g\ Al$$

shows that three conversions are required.

$$600.0g\ Al_2S_3 \times \frac{1\ mol\ Al_2S_3}{150.1g\ Al_2S_3} = 3.997\ mol\ Al_2S_3$$

$$3.997\ mol\ Al_2S_3 \times \frac{2\ mol\ Al}{1\ mol\ Al_2S_3} = 7.995\ mol\ Al$$

$$7.995\ mol\ Al \times \frac{26.98g\ Al}{1\ mol\ Al} = 215.7g\ Al$$

The individual steps can be combined as follows:

$$600.0g\ Al_2S_3 \times \frac{1\ mol\ Al_2S_3}{150.1g\ Al_2S_3} \times \frac{2\ mol\ Al}{1\ mol\ Al_2S_3} \times \frac{26.98g\ Al}{1\ mol\ Al} = 215.7g\ Al$$

***Practice Problem A

When chlorine gas is added to an aqueous solution of sodium bromide, liquid bromine and aqueous sodium chloride are produced.  How many grams of bromine are formed from 25.0g of chlorine?

29.   Ethane, $C_2H_{6(g)}$, burns in excess oxygen to produce carbon dioxide and water vapor.  If 72.0g of ethane are burned and 105g of carbon dioxide are formed, what is the percentage yield of carbon dioxide.

Find the theoretical yield of carbon dioxide from the balanced equation.

Step 1:   Write the initial equation.

$$C_2H_{6(g)} + O_{2(g)} \longrightarrow CO_{2(g)} + H_2O_{(g)}$$

Step 2:   Count the atoms of each element.

| element | reactants | products |
|---------|-----------|----------|
| C | 2 | 1 |
| H | 6 | 2 |
| O | 2 | 3 |

Step 3:   Insert coefficients for carbon.

$$C_2H_{6(g)} + O_{2(g)} \longrightarrow 2CO_{2(g)} + H_2O_{(g)}$$

Insert coefficients for hydrogen.

$$C_2H_{6(g)} + O_{2(g)} \longrightarrow 2CO_{2(g)} + 3H_2O_{(g)}$$

Insert coefficients for oxygen.  The coefficient $3\frac{1}{2}$ will balance oxygen atoms.

$$C_2H_{6(g)} + 3\tfrac{1}{2}O_{2(g)} \longrightarrow 2CO_{2(g)} + 3H_2O_{(g)}$$

To obtain whole-number coefficients, multiply the entire equation by 2.

$$2C_2H_{6(g)} + 7O_{2(g)} \longrightarrow 4CO_{2(g)} + 6H_2O_{(g)}$$

Step 4:   Recount atoms.

| element | reactants | products |
|---------|-----------|----------|
| C | 4 | 4 |
| H | 12 | 12 |
| O | 14 | 14 |

Step 5:   No change in coefficients is necessary.

Label the balanced equation with the given and required quantities.

$$2C_2H_{6(g)} + 7O_{2(g)} \longrightarrow 4CO_{2(g)} + 6H_2O_{(g)}$$

72.0g                                                       ?g (theoretical)
                                                        105g (actual)

Notice that the theoretical yield of carbon dioxide is unknown. The value, 105g, given in the problem refers to the actual yield.  The mole map,

$$72.0g\ C_2H_6 \longrightarrow ?\ mol\ C_2H_6 \longrightarrow ?\ mol\ CO_2 \longrightarrow ?\ g\ CO_2,$$

shows that three conversions are necessary.

$$72.0g\ C_2H_6 \times \frac{1\ mol\ C_2H_6}{30.06g\ C_2H_6} = 2.40\ mol\ C_2H_6$$

$$2.40\ mol\ C_2H_6 \times \frac{4\ mol\ CO_2}{2\ mol\ C_2H_6} = 4.80\ mol\ CO_2$$

$$4.80\ mol\ CO_2 \times \frac{44.0g\ CO_2}{1\ mol\ CO_2} = 211g\ CO_2$$

The individual steps can be combined as follows:

$$72.0g\ C_2H_6 \times \frac{1\ mol\ C_2H_6}{30.06g\ C_2H_6} \times \frac{4\ mol\ CO_2}{2\ mol\ C_2H_6} \times \frac{44.0g\ CO_2}{1\ mol\ CO_2} = 211g\ CO_2$$

The percentage yield is calculated using the following formula,

$$\%\ yield = \frac{actual\ yield}{theoretical\ yield} \times 100\%$$

$$= \frac{105g}{211g} \times 100\%$$

$$= 49.8\%$$

***Practice Problem B

Ammonia ($NH_3$) can react with oxygen to produce NO and water. When 26.0g $NH_3$ reacts, 35.7g of NO is produced.  What is the percentage yield of NO?

31.    If 25.0g of $SO_2$ and 6.00g of $O_2$ are allowed to react according to the following equation,

$$2SO_{2(g)} + O_{2(g)} \longrightarrow 2SO_{3(g)}$$

what is the limiting reactant?

Label the equation with the given quantities.

$$2SO_{2(g)} + O_{2(g)} \longrightarrow 2SO_{3(g)}$$
$$25.0g \qquad 6.00g$$

To find the limiting reactant, calculate the moles of $SO_3$ produced by each reactant.

$SO_2$:   25.0g $SO_2 \longrightarrow$ ? mol $SO_2 \longrightarrow$ ? mol $SO_3$

$O_2$:   6.00g $O_2 \longrightarrow$ ? mol $O_2 \longrightarrow$ ? mol $SO_3$

Step 1:   Calculate the moles of each reactant.

$SO_2$:   $25.0g\ SO_2 \times \dfrac{1\ mol\ SO_2}{64.1g\ SO_2} = 0.390\ mol\ SO_2$

$O_2$:   $6.00g\ O_2 \times \dfrac{1\ mol\ O_2}{32.0g\ O_2} = 0.188\ mol\ O_2$

Step 2:   Calculate the moles of $SO_3$ that would be produced by each reactant.

$SO_2$:   $0.390\ mol\ SO_2 \times \dfrac{2\ mol\ SO_3}{2\ mol\ SO_2} = 0.390\ mol\ SO_3$

$O_2$:   $0.188\ mol\ O_2 \times \dfrac{2\ mol\ SO_3}{1\ mol\ O_2} = 0.376\ mol\ SO_3$

Since $O_2$ produces the lower theoretical yield, $O_2$ is the limiting reactant.

***Practice Problem C

Carbon dioxide reacts with water to produce carbonic acid, $H_2CO_3$. What weight of carbonic acid can be produced from 15.5g $CO_2$ and 10.0g of water?

33.    When 3.50g of $NaNH_2$ and 3.50g of $NaNO_3$ were allowed to react according to the following equation,

$$3NaNH_{2(S)} + NaNO_{3(S)} \longrightarrow NaN_{3(S)} + 3NaOH_{(S)} + NH_{3(g)}$$

1.20g of $NaN_3$ was formed.  What was the percentage yield of $NaN_3$?

Label the equation with the given and required quantities.

$$3NaNH_{2(S)} + NaNO_{3(S)} \longrightarrow NaN_{3(S)} + 3NaOH_{(S)} + NH_{3(g)}$$

3.50g          3.50g              ? g

This is a limiting reactant problem since the mass of two reactants is given.  Calculate the moles of $NaN_3$ produced by each reactant.

$NaNH_2$:  The mole map,

3.50g $NaNH_2$ $\longrightarrow$ ? mol $NaNH_2$ $\longrightarrow$ ? mol $NaN_3$

shows that two conversions are necessary.

$$3.50g\ NaNH_2 \times \frac{1\ mol\ NaNH_2}{39.0g\ NaNH_2} \times \frac{1\ mol\ NaN_3}{3\ mol\ NaNH_2} = 0.0299\ mol\ NaN_3$$

$NaNO_3$:  The mole map,

3.50g $NaNO_3$ $\longrightarrow$ ? mol $NaNO_3$ $\longrightarrow$ ? mol $NaN_3$

shows that two conversions are necessary.

$$3.50g\ NaNO_3 \times \frac{1\ mol\ NaNO_3}{85.0g\ NaNO_3} \times \frac{1\ mol\ NaN_3}{1\ mol\ NaNO_3} = 0.0412\ mol\ NaN_3$$

Since $NaNH_2$ produces the lower theoretical yield of $NaN_3$, $NaNH_2$ is the limiting reactant.

The theoretical yield of $NaN_3$ in grams must be calculated to find percentage yield.

$$0.0299\ mol\ NaN_3 \times \frac{65.0g\ NaN_3}{1\ mol\ NaN_3} = 1.94g\ NaN_3$$

$$\%\ yield = \frac{actual\ yield}{theoretical\ yield} \times 100\%$$

$$= \frac{1.20g\ NaN_3}{1.94g\ NaN_3}$$

$$= 61.9\%$$

***Practice Problem D

A mixture of 20.0g of magnesium reacts with 25.0g of chlorine. If 30.0g of magnesium chloride is produced, what is the percentage yield of magnesium chloride?

---

## SELF-TEST

Circle the correct answer in the following multiple choice questions.

1.  For the reaction,

$$2Mg_{(l)} + TiCl_{4\,(g)} \longrightarrow 2MgCl_{2\,(s)} + Ti_{(s)}$$

a.  magnesium is a product.
b.  magnesium is a solid.
c.  two magnesium atoms react with one $TiCl_4$ formula unit.
d.  two grams of magnesium reacts with one gram of $TiCl_4$.

2.  When 50.00g of potassium chlorate decomposes by the reaction,

$$2KClO_3 \xrightarrow{\Delta} 2KCl + 3O_2;$$ the mass of the products are

a.  50.00g KCl and 75.00g $O_2$
b.  30.42g KCl and 19.58g $O_2$
c.  25.00g KCl and 25.00g $O_2$
d   20.00g KCl and 30.00g $O_2$

3.  Which of the following equations is not balanced?

a.  $4Li_{(s)} + O_{2\,(g)} \longrightarrow 2Li_2O$

b.  $2AgNO_{3\,(aq)} + MgCl_{2\,(aq)} \longrightarrow 2AgCl_{(s)} + Mg(NO_3)_{2\,(aq)}$

c. $C_3H_{8(g)} + 5O_{2(g)} \longrightarrow 3CO_{2(g)} + 4H_2O_{(g)}$

d. $2HCl_{(aq)} + Mg(OH)_2 \longrightarrow H_2O_{(l)} + MgCl_{2(aq)}$

4. The reaction $Cu_{(s)} + 2AgNO_{3(aq)} \longrightarrow Cu(NO_3)_{2(aq)} + 2Ag_{(s)}$ is classified as

a. decomposition reaction
b. combination reaction
c. single-replacement reaction
d. double-replacement reaction

5. How many moles of water can be produced from 3.0 moles of ammonia ($NH_3$) in the following reaction?

$$4NH_{3(g)} + 5O_{2(g)} \longrightarrow 4NO_{(g)} + 6H_2O_{(g)}$$

a. 4.5 moles $H_2O$

b. 2.0 moles $H_2O$

c. 6.0 moles $H_2O$

d. 5.0 moles $H_2O$

6. What weight of iron will react with 25.00g of oxygen in the following reaction?

$$4Fe_{(s)} + 3O_{2(g)} \longrightarrow 2Fe_2O_{3(s)}$$

a. 32.72g Fe
b. 43.63g Fe
c. 58.18g Fe
d. 116.4g Fe

7. How many moles of sodium chloride can be produced from 0.75 moles of sodium and 0.40 moles of chlorine?

$$2\,Na_{(s)} + Cl_{2(g)} \longrightarrow 2NaCl_{(s)}$$

a. 0.75 mol NaCl
b. 0.40 mol NaCl
c. 0.80 mol NaCl
d. 1.15 mol NaCl

8. When 10.0g of carbon dioxide reacts with an excess of sodium oxide, 10.0g of sodium carbonate is recovered.

$$Na_2O_{(s)} + CO_{2(g)} \longrightarrow Na_2CO_{3(s)}$$

What is the percentage yield?

a. 100%
b. 41.5%
c. 24.1%
d. 10.0%

9.    Potassium oxide, $K_2O$, reacts with water to form potassium
      hydroxide, KOH.  How many moles of KOH can be produced from
      25.0g of $K_2O$ and 5.00g of water?

      a.   .133 mol KOH
      b.   .139 mol KOH
      c.   .531 mol KOH
      d.   .556 mol KOH

10.   When 10.0g of water and 45.5g of dinitrogen pentoxide, $N_2O_5$,
      react, 40.0g of nitric acid is formed.  What is the percentage
      yield?

      a.   57.1%
      b.   28.6%
      c.   37.7%
      d.   75.3%

---

## ANSWERS TO PRACTICE PROBLEMS

Practice Problem A

$$Cl_{2\,(g)} + 2NaBr_{(aq)} \longrightarrow Br_{2\,(l)} + 2NaCl_{(aq)}$$

25.0g                    ? g

The mole map,

$$25.0\text{g } Cl_2 \longrightarrow \text{? mol } Cl_2 \longrightarrow \text{? mol } Br_2 \longrightarrow \text{? g } Br_2$$

shows that three conversions are required.

$$25.0\text{g } Cl_2 \times \frac{1 \text{ mol } Cl_2}{70.9\text{g } Cl_2} = 0.353 \text{ mol } Cl_2$$

$$0.353 \text{ mol } Cl_2 \times \frac{1 \text{ mol } Br_2}{1 \text{ mol } Cl_2} = 0.353 \text{ mol } Br_2$$

$$0.353 \text{ mol } Br_2 \times \frac{159.8\text{g } Br_2}{1 \text{ mol } Br_2} = 56.4\text{g } Br_2$$

The individual steps can be combined as follows:

$$25.0\text{g } Cl_2 \times \frac{1 \text{ mol } Cl_2}{70.9\text{g } Cl_2} \times \frac{1 \text{ mol } Br_2}{1 \text{ mol } Cl_2} \times \frac{159.8\text{g } Br_2}{1 \text{ mol } Br_2} = 56.3\text{g } Br_2$$

The answers differ due to the rounding off of intermediate
answers in the first method.

Practice Problem B

$$4NH_3 + 5O_2 \longrightarrow 4 \text{ NO} + 6 \text{ H}_2O$$

26.0g                    ? g (theoretical)
                         35.7g (actual)

The mole map,

$$26.0\text{g } NH_3 \longrightarrow \text{? mol } NH_3 \longrightarrow \text{? mol NO} \longrightarrow \text{? g NO}$$

shows that three conversions are necessary to calculate the
theoretical yield.

$$26.0g \; NH_3 \times \frac{1 \; mol \; NH_3}{17.0g \; NH_3} = 1.53 \; mol \; NH_3$$

$$1.53 \; mol \; NH_3 \times \frac{4 \; mol \; NO}{4 \; mol \; NH_3} = 1.53 \; mol \; NO$$

$$1.53 \; mol \; NO \times \frac{30.0g \; NO}{1 \; mol \; NO} = 45.9g \; NO$$

In a single step,

$$26.0g \; NH_3 \times \frac{1 \; mol \; NH_3}{17.0g \; NH_3} \times \frac{4 \; mol \; NO}{4 \; mol \; NH_3} \times \frac{30.0g \; NO}{1 \; mol \; NO} = 45.9g \; NO$$

$$\% \; yield = \frac{actual \; yield}{theoretical \; yield} \times 100\%$$

$$= \frac{35.7g}{45.9g} \times 100\% = 77.8\%$$

Practice Problem C

$$CO_{2 \, (g)} + H_2O_{(g)} \longrightarrow H_2CO_{3 \, (aq)}$$

$$15.5g \qquad 10.0g \qquad \qquad ? \; g$$

To find the limiting reactant, calculate the moles of $H_2CO_3$ produced by each reactant.

$$CO_2: \quad 15.5g \; CO_2 \longrightarrow ? \; mol \; CO_2 \longrightarrow ? \; mol \; H_2CO_3$$

$$15.5g \; CO_2 \times \frac{1 \; mol \; CO_2}{44.0g \; CO_2} = 0.352 \; mol \; CO_2$$

$$0.352 \; mol \; CO_2 \times \frac{1 \; mol \; H_2CO_3}{1 \; mol \; CO_2} = 0.352 \; mol \; H_2CO_3$$

$$H_2O: \quad 10.0g \; H_2O \longrightarrow mol \; H_2O \longrightarrow ? \; mol \; H_2CO_3$$

$$10.0g \; H_2O \times \frac{1 \; mol \; H_2O}{18.0g \; H_2O} = 0.556 \; mol \; H_2O$$

$$0.556 \; mol \; H_2O \times \frac{1 \; mol \; H_2CO_3}{1 \; mol \; H_2O} = 0.556 \; mol \; H_2CO_3$$

Since $CO_2$ produces the lower theoretical yield, $CO_2$ is the limiting reactant.  To find the mass of carbonic acid, convert 0.352 mol $H_2CO_3$ to grams.

$$0.352 \; mol \; H_2CO_3 \times \frac{62.0g \; H_2CO_3}{1 \; mol \; H_2CO_3} = 21.8g \; H_2CO_3$$

Practice Problem D

$$Mg \quad + \quad Cl_2 \quad \longrightarrow \quad MgCl_2$$

20.0g    25.0g      ? g (theoretical)
                              30.0g (actual)

Calculate the moles of $MgCl_2$ produced by each reactant.

Mg:   20.0g Mg $\longrightarrow$ ? mol Mg $\longrightarrow$ ? mol $MgCl_2$

$$20.0g\ Mg \times \frac{1\ mol\ Mg}{24.3g\ Mg} \times \frac{1\ mol\ MgCl_2}{1\ mol\ Mg} = 0.823\ mol\ MgCl_2$$

$Cl_2$:   25.0g $Cl_2$ $\longrightarrow$ ? mol $Cl_2$ $\longrightarrow$ ? mol $MgCl_2$

$$25.0g\ Cl_2 \times \frac{1\ mol\ Cl_2}{70.9g\ Cl_2} \times \frac{1\ mol\ MgCl_2}{1\ mol\ Cl_2} = 0.353\ mol\ MgCl_2$$

Since $Cl_2$ produces the lower theoretical yield, $Cl_2$ is the limiting reactant.  Find the mass of 0.353 mol $MgCl_2$.

$$0.353\ mol\ MgCl_2 \times \frac{95.2g\ MgCl_2}{1\ mol\ MgCl_2} = 33.6g\ MgCl_2$$

$$\%\ yield = \frac{actual\ yield}{theoretical\ yield} \times 100\%$$

$$= \frac{30.0g}{33.6g} \times 100\% = 89.3\%$$

---

## ANSWERS TO SELF-TEST

1.   c
2.   b
3.   d
4.   c
5.   a
6.   c
7.   a
8.   b
9.   c
10.   d

# GASES

CHAPTER 10

---

## SUMMARY OUTLINE

### 10.1 Properties of Gases

The gaseous state of matter, like all matter, has mass and occupies space.

Gases exert pressure. Pressure is a force applied on a surface, or unit area. Gas pressure is measured with a barometer and expressed in units of psi, mm Hg, torr, atmosphere (atm), and pascal (Pa).

$$1 \text{ mm Hg} = 1 \text{ torr}$$
$$1 \text{ atm} = 760 \text{ torr}$$
$$1 \text{ atm} = 1.01 \times 10^5 \text{ Pa}$$

Gases can be compressed. Gases expand when heated and contract when cooled.

Gases diffuse through each other to form homogeneous mixtures.

### 10.2 Volume and Pressure Relationships

There are four variables used in calculations involving gases: mass, volume, pressure, and temperature. Two variables are held constant while the relationship between the other two is studied.

Boyle's law states that the pressure and volume of a gas are inversely related; that is, an increase in pressure causes a decrease in volume, and vice versa.

$$P \propto \frac{1}{V}; \qquad \text{or} \quad PV = \text{constant}$$

$$P_i V_i = P_f V_f \quad \text{or} \quad V_f = V_i \times P_{ratio}$$

$$P_f = P_i \times V_{ratio}$$

## 10.3   Volume and Temperature Relationships

Charles' law states that the volume and absolute temperature of a gas are directly related; that is, an increase in absolute temperature causes an increase in volume, and vice versa.

$$V \propto T; \quad \text{or}$$

$$\frac{V}{T} = \text{constant}$$

$$\frac{V_i}{T_i} = \frac{V_f}{T_f} \quad \text{or} \quad \begin{aligned} V_f &= V_i \times T_{ratio} \\ T_f &= T_i \times V_{ratio} \end{aligned}$$

On the Kelvin temperature scale, 0 K (absolute zero) represents the temperature at which a gas would have no volume and no kinetic energy.

$$K = {}^\circ C + 273$$

## 10.4   Pressure and Temperature Relationships

The pressure exerted by a gas at constant volume is directly proportional to absolute temperature.

$$P \propto T; \quad \text{or} \qquad \frac{P}{T} = \text{constant}$$

$$\frac{P_i}{T_i} = \frac{P_f}{T_f} \quad \text{or} \qquad \begin{aligned} P_f &= P_i \times T_{ratio} \\ T_f &= T_i \times P_{ratio} \end{aligned}$$

## 10.5   Combining the Gas Laws

$$\frac{P_i V_i}{T_i} = \frac{P_f V_f}{T_f} \quad \text{or}$$

$$V_f = V_i \times P_{ratio} \times T_{ratio}$$

Since pressure is inversely proportional to volume, when pressure increases, $P_{ratio} < 1$; when pressure decreases, $P_{ratio} > 1$. Temperature is directly proportional to volume. When temperature increases, $T_{ratio} > 1$; when temperature decreases, $T_{ratio} < 1$.

$$P_f = P_i \times V_{ratio} \times T_{ratio}$$

$$T_f = T_i \times V_{ratio} \times P_{ratio}$$

The arrangement of the measurements in each ratio is determined by the relationship between the variables.

## 10.6  Avogadro's Hypothesis

Avogadro's hypothesis states that at the same conditions of temperature and pressure, equal volumes of gases contain equal number of molecules.

Standard temperature and pressure (STP) is defined as 1 atm and 273 K (0°C).

One mole of any ideal gas occupies 22.4L at STP.  This volume is called the molar volume.

## 10.7  The Ideal Gas Equation

The gas laws and Avogadro's hypothesis can all be expressed in terms of volume,

Boyle's law:     $V \propto \dfrac{1}{p}$     T, n constant

Charles' law:    $V \propto T$     P, n constant

Avogadro's
hypothesis:     $V \propto n$     P, T constant

where n = number of moles of the gas.

When these three relationships are combined;

$$V \propto \frac{nT}{p}$$

$$V = \frac{nRT}{P} \text{ or}$$

$$PV = nRT \text{ (ideal gas equation)}$$

R, the molar gas constant, is equal to 0.0821 $\dfrac{L \cdot atm}{mol \cdot K}$.

## 10.8  The Kinetic Molecular Theory

The kinetic molecular theory is based on Avogadro's hypothesis.

1. All gases consist of separate and distinct particles.
2. A gas molecule has negligible volume.
3. Gas molecules are very far apart.  The volume occupied by a gas is mostly empty space.
4. Gas molecules move in random, straight-line motion.
5. Collisions between gas molecules are elastic.
6. At a given temperature, the average kinetic energy of the molecules of two different gases is the same.

## 10.9  Gas Mixtures

Dalton's law of partial pressures states that each gas in a mixture of gases exerts a pressure independent of the other gases.

$$P_{total} = P_1 + P_2 + P_3 + \ldots$$

where $P_{total}$ is the total pressure of the mixture, and $P_1$, $P_2$, $P_3$, ... are the partial pressures of the individual gases.

## 10.10  Gas Density

Gases have low densities relative to solids and liquids. The density of a gas is dependent on temperature and pressure.

## 10.11  Stoichiometric Calculations Involving Gas Volumes (Optional)

There are three categories of stoichiometric calculations involving gaseous volume:  mole-volume calculations, weight-volume calculations, and volume-volume calculations.

The fundamental relationships important to gas volume stoichiometric calculations are summarized in the map below:

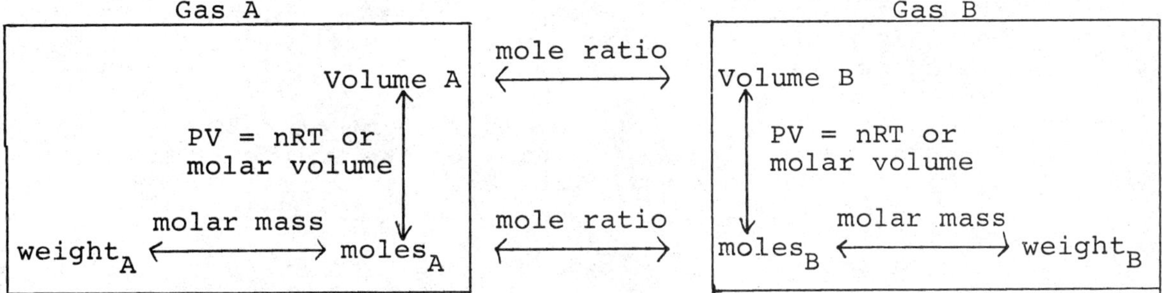


Mole-volume calculations involve the relationship 1 mole = 22.4L when the conditions are at STP.  At all other conditions, the ideal gas equation must be used.

---

SOLUTIONS TO SELECTED STUDY QUESTIONS AND PROBLEMS
and
PRACTICE PROBLEMS

---

10.    A sample of gas has a volume of 360 mL at a pressure of 0.750 atm.  If the temperature is constant, what volume in mL will the gas occupy at 1.00 atm?  State the problem:

$$V_i = 360 \text{ mL} \qquad V_f = ?$$
$$P_i = 0.750 \text{ atm} \qquad P_f = 1.00 \text{ atm}$$
$$T_i = \text{constant} \qquad T_f = \text{constant}$$

$V_f$ is designated as the unknown variable.  The value for $V_f$ will be related to the value of $V_i$, the known variable, multiplied by a pressure (P) and a temperature (T) ratio.

$$V_f = V_i \times P_{ratio} \times T_{ratio}$$

Since T is constant, the $T_{ratio}$ term equals one.  The solution can be simplified to,

$$V_f = V_i \times P_{ratio}$$

$$V_f = 360 \text{ mL} \times \frac{?\ \text{atm}}{?\ \text{atm}}$$

To determine $P_{ratio}$, determine how P changes and what effect this change will have on V.  Since the pressure increases, the volume must decrease, and $P_{ratio} < 1$.  (Remember that P and V are inversely related).

$$V_f = 360 \text{ mL} \times \frac{0.750 \ \cancel{\text{atm}}}{1.00 \ \cancel{\text{atm}}}$$

$$V_f = 270 \text{ mL}$$

As a check, compare $V_f$ to $V_i$.  $V_f < V_i$ as predicted.

Whenever initial and final conditions are given for V, P, and T in a problem, a known variable will be multiplied by ratios of the other variables.

18.  A sample of gas has a volume of 79.5 mL at 45°C.  If the pressure is constant, what will be the volume of the sample in mL at 0°C?  State the problem:

| | |
|---|---|
| $V_i = 79.5$ mL | $V_f = ?$ |
| $P_i =$ constant | $P_f =$ constant |
| $T_i = 45°C + 273 = 318$ K | $T_f = 0°C + 273 = 273$K |

Remember that temperatures in any gas law calculation must be expressed in kelvins.

$$V_f = V_i \times P_{ratio} \times T_{ratio}$$

Since P is constant, the $P_{ratio}$ term equals one.

$$V_f = V_i \times T_{ratio}$$

$$V_f = 79.5 \text{ mL} \times \frac{?\ \text{K}}{?\ \text{K}}$$

Since the temperature decreases, the volume must decrease, and $T_{ratio} < 1$.  (Remember that T and V are directly related).

$$V_f = 79.5 \text{ mL} \times \frac{273 \ \cancel{\text{K}}}{318 \ \cancel{\text{K}}}$$

$$V_f = 68.3 \text{ mL}$$

As a check, compare $V_f$ to $V_i$.  $V_f < V_i$ as predicted.

26.  A gas sample was heated from −5.0°C to 90.0°C, and the volume increased from 1.00 L to 3.00 L.  If the initial pressure was 0.800 atm, what was the final pressure in atm?

State the problem.

$V_i$ = 1.00 L                    $V_f$ = 3.00 L

$P_i$ = 0.800 atm                 $P_f$ = ?

$T_i$ = -5.0°C + 273 = 268K       $T_f$ = 90.0°C + 273 = 363K

$$P_f = P_i \times V_{ratio} \times T_{ratio}$$

$$P_f = 0.800 \text{ atm} \times V_{ratio} \times T_{ratio}$$

Since the volume increases, the pressure must decrease and $V_{ratio}$ <1.

$$P_f = 0.800 \text{ atm} \times \frac{1.00 \text{ L}}{3.00 \text{ L}} \times T_{ratio}$$

Since the temperature increases the pressure must increase and $T_{ratio}$ >1.

$$P_f = 0.800 \text{ atm} \times \frac{1.00 \text{ L}}{3.00 \text{ L}} \times \frac{363 \text{ K}}{268 \text{ K}}$$

$$P_f = 0.361 \text{ atm}$$

The V ratio decreases $P_f$ and the $T_{ratio}$ increases $P_f$.  When these opposing changes occur, the value for $P_f$ cannot be checked against the value for $P_i$.

Sections 10.5 and 10.7 in the text refer to the following equation,

$$\frac{P_i V_i}{T_i} = \frac{P_f V_f}{T_f}$$

Rearrangement of this equation to solve for $P_f$ results in the following equation,

$$P_f = \frac{P_i V_i T_f}{T_i V_f} = P_i \times \frac{V_i}{V_f} \times \frac{T_f}{T_i}$$

$$= 0.800 \text{ atm} \times \frac{1.00 \text{ L}}{3.00 \text{ L}} \times \frac{363 \text{ K}}{268 \text{ K}}$$

$$P_f = 0.361 \text{ atm}$$

Both methods of solving "changing conditions" gas law problems result in the same solution.  The first method relies on a knowledge of relationships between P, V, and T and the second method relies on a knowledge of the combined gas law equation and algebraic rearrangements of this equation.

***Practice Problem A

Find the unknown variable for the following gases:

a.  A sample of methane gas ($CH_4$) has a volume of 5.75 L at a temperature of 22.5°C and a pressure of 755 torr. What volume will the gas occupy at 740 torr if the temperature is held constant?

b.  $O_2$ gas, at a temperature of 21.0°C and a pressure of 645 torr, has a volume of 25.0 mL.  If the volume is halved and the temperature increases by 10.0°C, what is the pressure of the gas?

32.  What is the volume of $4.16 \times 10^{20}$ molecules of carbon monoxide at STP?

$$4.16 \times 10^{20} \text{ molecules CO} \longrightarrow \text{? mol CO} \longrightarrow \text{? L CO}$$

To convert from mol CO to L CO, remember that 1 mole of any gas occupies 22.4 L at STP.

$$4.16 \times 10^{20} \text{ molecules CO} \xrightarrow{6.02 \times 10^{23}} \text{? mol CO} \xrightarrow{\text{molar volume}} \text{L CO}$$

$$4.16 \times 10^{20} \text{ molecules CO} \times \frac{1 \text{ mol CO}}{6.02 \times 10^{23} \text{ molecules}} \times \frac{22.4 \text{ L CO}}{1 \text{ mol CO}} =$$

$$1.55 \times 10^{-2} \text{ L CO}$$

***Practice Problem B

What is the mass of 10.0 L of chlorine gas at STP?

38.     What volume will 10.0g of $CO_2$ have at 25°C and 1.75 atm?

The ideal gas equation is used when there is no change in variables.

$$PV = nRT$$

In this problem, the volume is required, so the equation is rearranged.

$$V = \frac{nRT}{P}$$

The quantity of gas must be in moles.  Convert 10.0g of $CO_2$ to moles:

$$10.0\text{g } CO_2 \times \frac{1 \text{ mol } CO2}{44.0\text{g } CO_2} = 0.227 \text{ mol } CO_2$$

The temperature must be in kelvins,

$$T = 25°C + 273 = 298 \text{ K}$$

Substitute numerical values for n, R, T and P.

$$V = \frac{0.227 \text{ mol } O_2 \times 0.0821 \frac{L \cdot atm}{mol \cdot K} \times 298 \text{ K}}{1.75 \text{ atm}}$$

$$= 3.17 \text{ L}$$

*** Practice Problem C

What mass of nitrogen gas occupies 20.5 L at 22.0°C and 750 torr?

50.    Calculate the density of each gas at STP.

   a.   $CO_2$

   At STP, one mole of any gas occupies 22.4 L.  If the conditions given are not at STP, the ideal gas equation can be used to find the volume of 1 mole of the gas.  The mass of one mole of $CO_2$ is found from the atomic weights.

$$FW\ CO_2 = (1\ \text{atom C})\ \frac{12.0\ amu}{1\ \text{atom C}} + (2\ \text{atoms O})\ \frac{16.0\ amu}{1\ \text{atom O}}$$

$$= 12.0\ amu + 32.0\ amu$$

$$= 44.0\ amu$$

The molar mass of $CO_2$ is 44.0g.

$$d = \frac{m}{v}$$

$$d = \frac{44.0g}{22.4\ L} = 1.96\ \frac{g}{L}$$

Consult Appendix C in the text for answers to parts b and c.

***Practice Problem D

What is the density of $CO_2$ at 25°C and 2.0 atm?

54.    In the complete combustion of octane ($C_8H_{18}$),

$$2C_8H_{18\ (g)} + 25\ O_{2\ (g)} \longrightarrow 16\ CO_{2\ (g)} + 18\ H_2O_{(g)}$$

what volume of $CO_2$ is produced from 0.750g of octane at 400.0°C and 10.0 atm?

Label the balanced chemical equation with the given and required quantities.

$$2C_8H_{18\ (g)} + 25\ O_{2\ (g)} \longrightarrow 16\ CO_{2\ (g)} + 18\ H_2O_{(g)}$$

0.750g                                          ? L
                                    400.0°C + 273 = 673 K
                                    10.0 atm

The map in section 10.11 illustrates the following solution:

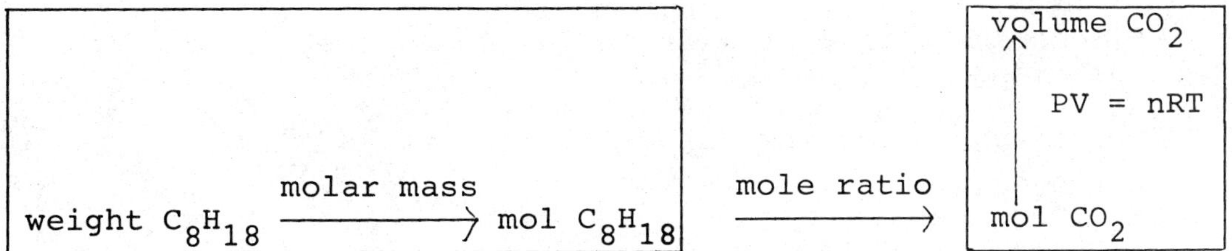

The solution has two distinct parts, a stoichiometric calculation (weight $C_8H_{18}$ $\longrightarrow$ mol $CO_2$) and an ideal gas law calculation (mol $CO_2$ $\longrightarrow$ volume $CO_2$).  The stoichiometric calculation involves two conversions.

$$0.750g\ C_8H_{18} \times \frac{1\ mol\ C_8H_{18}}{114g\ C_8H_{18}} = 6.58 \times 10^{-3}\ mol\ C_8H_{18}$$

$$6.58 \times 10^{-3}\ mol\ C_8H_{18} \times \frac{16\ mol\ CO_2}{2\ mol\ C_8H_{18}} = 5.26 \times 10^{-2}\ mol\ CO_2$$

This molar quantity of $CO_2$ is used in the ideal gas equation to find volume.

$$PV = nRT$$

$$V = \frac{nRT}{P} = \frac{5.26 \times 10^{-2}\ mol\ CO_2 \times 0.0821\ \frac{L \cdot atm}{mol \cdot K} \times 673\ K}{10.0\ atm}$$

$$V = 0.291\ L$$

*** Practice Problem E

Potassium chlorate decomposes when heated to form potassium chloride and oxygen.

$$2\ KClO_{3(S)} \xrightarrow{\Delta} 2\ KCl_{(S)} + 3\ O_{2(g)}$$

What mass of potassium chlorate must decompose to produce 0.500 L of oxygen gas at 30°C and 0.948 atm?

---

## SELF-TEST

Circle the correct answer in the following multiple choice questions.

1. The pressure 745 torr is equivalent to

   a. 0.745 mm Hg
   b. 1.02 atm
   c. 0.980 atm
   d. $7.45 \times 10^3$ mm Hg

2. Which of the following statements is not true for gases?

   a. Gases exert pressure.
   b. Gases are compressible.
   c. Gases expand when cooled.
   d. Gases diffuse through each other.

3. Which of the following solutions will correctly find $V_f$ if $V_i = 25.0$ ml, $P_i = 1.00$ atm, $P_f = 1.50$ atm, $T_i = 22.0°C$ and $T_f = 29.0°C$.

   a. $V_f = 25.0$ ml $\times \dfrac{1.00 \text{ atm}}{1.50 \text{ atm}} \times \dfrac{29.0°C}{22.0°C}$

   b. $V_f = 25.0$ ml $\times \dfrac{1.50 \text{ atm}}{1.00 \text{ atm}} \times \dfrac{302 \text{ K}}{295 \text{ K}}$

   c. $V_f = 25.0$ ml $\times \dfrac{1.50 \text{ atm}}{1.00 \text{ atm}} \times \dfrac{295 \text{ K}}{302 \text{ K}}$

   d. $V_f = 25.0$ ml $\times \dfrac{1.00 \text{ atm}}{1.50 \text{ atm}} \times \dfrac{302 \text{ K}}{295 \text{ K}}$

4. Which of the following statements is true?

   a. The standard conditions of temperature and pressure are 25°C and 1 atm.
   b. The volume of a mole of gas at STP is 2.24 L.
   c. 70.9g of $CL_2$ at 0°C and 1 atm will occupy 22.4 L.
   d. The standard conditions of temperature and pressure are 0 K and 1 atm.

5. For the reaction,

   $$3\ C_3H_{8\,(g)} + 5\ O_{2\,(g)} \longrightarrow 3\ CO_{2\,(g)} + 4\ H_2O_{(g)}$$

   which of the following statements is true?

   a. 3 liters of $C_3H_8$ will produce 3 liters of $CO_2$ if enough oxygen is present.
   b. Since $C_3H_8$ and $O_2$ are present at the same temperature and pressure, there are an equal number of $C_3H_8$ and $O_2$ molecules.
   c. At STP 3 moles of $C_3H_8$ will occupy 22.4 L.
   d. The law of conservation of matter does not apply to gases, since 8 moles of reactant gases yield only 7 moles of product gases.

6.     The volume of 2.00 moles of $H_2$ at 740 torr and 22.0°C is

   a.   $4.88 \times 10^{-3}$ L
   b.   $6.55 \times 10^{-2}$ L
   c.   49.7 L
   d.   $2.01 \times 10^{-2}$ L

7.     A sample of He has a volume of 3.0 L at 27°C.  At what
       temperature would this sample have a volume of 6.0 L?

   a.   54°C
   b.   600°C
   c.   327 K
   d.   327°C

8.     Which of the following gas samples would have the greatest
       average kinetic energy?

   a.   $H_2$ at 25°C
   b.   $I_2$ at 25°C
   c.   $UF_6$ at 0°C
   d.   $O_2$ at 50°C

9.     The density of $SO_2$ at 20°C and 750 torr is

   a.   $3.80 \times 10^{-1}$ g/L
   b.   2.62 g/L
   c.   38.6 g/L
   d.   2.86 g/L

10.    What volume of $O_2$ can be prepared by the electrolysis of
       1.00 mole of water at 22.0°C and 1 atm?

$$2 \, H_2O_{(g)} \longrightarrow 2 \, H_2{}_{(g)} + O_2{}_{(g)}$$

   a.   12.1 L
   b.   24.2 L
   c.   11.2 L
   d.   .903 L

---

## ANSWERS TO PRACTICE PROBLEMS

Practice Problem A

   a.   State the problem.

   $V_i = 5.75$ L               $V_f = ?$

   $P_i = 755$ torr             $P_f = 740$ torr

   $T_i = $ constant            $T_f = $ constant

   $$V_f = V_i \times P_{ratio} \times T_{ratio}{}^1$$

   $$V_f = 5.75 \text{ L} \times \frac{? \text{ torr}}{? \text{ torr}}$$

Since the pressure decreases, the volume must increase, and $P_{ratio} > 1$.

$$V_f = 5.75 \text{ L} \times \frac{755 \text{ torr}}{740 \text{ torr}}$$

$$V_f = 5.87 \text{ L}$$

$V_f > V_i$ as predicted.

b.  State the problem.

| | |
|---|---|
| $V_i = 25.0 \text{ mL}$ | $V_f = 12.5 \text{ mL}$ |
| $P_i = 645 \text{ torr}$ | $P_f = ?$ |
| $T_i = 21.0°C + 273 = 294 \text{ K}$ | $T_f = 304 \text{ K}$ |

$$P_f = P_i \times V_{ratio} \times T_{ratio}$$

$$P_f = 645 \text{ torr} \times V_{ratio} \times T_{ratio}$$

Since the volume decreases, the pressure must increase and $V_{ratio} > 1$.

$$P_f = 645 \text{ torr} \times \frac{25.0 \text{ mL}}{12.5 \text{ mL}} \times T_{ratio}$$

Since the temperature increases, the pressure must increase and $T_{ratio} > 1$.

$$P_f = 645 \text{ torr} \times \frac{25.0 \text{ mL}}{12.5 \text{ mL}} \times \frac{304 \text{ K}}{294 \text{ K}}$$

$$P_f = 1.33 \times 10^3 \text{ torr}$$

Practice Problem B

$$10.0 \text{ L CO} \longrightarrow ? \text{ mol CO} \longrightarrow ? \text{ g CO}$$

Since CO is measured at STP, the molar volume of CO can be used to find mol CO.

$$10.0 \text{ L CO} \xrightarrow{\text{molar volume}} ? \text{ mol CO} \xrightarrow{\text{molar mass}} ? \text{ g CO}$$

$$10.0 \text{ L CO} \times \frac{? \text{ mol CO}}{? \text{ L CO}} \times \frac{? \text{ g CL}}{? \text{ mol CO}} =$$

$$10.0 \text{ L CO} \times \frac{1 \text{ mol CO}}{22.4 \text{ L CO}} \times \frac{28.01 \text{g CO}}{1 \text{ mol CO}} = 12.5 \text{g CO}$$

Practice Problem C

State the problem.

$$V = 20.5 \text{ L}$$
$$T = 22.0°C + 273 = 295 \text{ K}$$
$$P = 750 \text{ torr} \times \frac{1 \text{ atm}}{760 \text{ torr}} = .987 \text{ atom}$$

mass = ?

There is no change in variables in the problem, so the ideal gas equation is used.  Rearrange the equation to solve for n (moles).   Once n is found, convert to mass.

$$n = \frac{PV}{RT} = \frac{.987 \text{ atm} \times 20.5 \text{ L}}{0.0821 \frac{\text{L·atm}}{\text{mol·K}} \times 295 \text{ K}}$$

$$= 0.835 \text{ mol } N_2$$

Notice that only temperature in kelvin, volume in liters, and pressure in atm will cancel the units in the gas constant R.

$$0.835 \text{ mol } N_2 \times \frac{28.02\text{g } N_2}{1 \text{ mol } N_2} = 23.4\text{g } N_2$$

## Practice Problem D

Find the volume of 1.00 mol of $CO_2$ (44.01g) using the ideal gas equation.

$$V = \frac{nRT}{P}$$

$$V = \frac{1.00 \text{ mol} \times 0.0821 \frac{\text{L·atm}}{\text{mol·K}} \times 298 \text{ K}}{2.0 \text{ atm}}$$

$$= 12 \text{ L}$$

Then find the density of $CO_2$.

$$d = \frac{m}{V} = \frac{44.01\text{g}}{12 \text{ L}} = 3.7 \frac{\text{g}}{\text{L}}$$

## Practice Problem E

$$2 \text{ KClO}_{3\,(S)} \xrightarrow{\Delta} 2 \text{ KCl}_{(S)} + 3 \text{ O}_{2\,(g)}$$

$$? \text{ g} \qquad\qquad\qquad\qquad \begin{array}{c} 0.500 \text{ L} \\ 30°C,\ 0.948 \text{ atm} \end{array}$$

The map in section 10.11 illustrated the following solution:

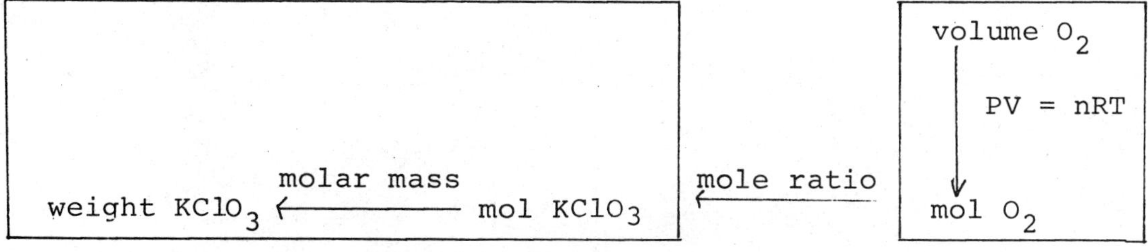


The first part of the solution involves an ideal gas calculation.

$$n_{O_2} = \frac{PV}{RT} = \frac{0.948 \text{ atm} \times 0.500 \text{ L}}{0.0821 \frac{\text{L·atm}}{\text{mol·K}} \times 303 \text{ K}}$$

$$= 1.91 \times 10^{-2} \text{ mol } O_2$$

Now use stoichiometric calculations to find weight of $KClO_3$.

$$1.91 \times 10^{-2} \text{ mol } O_2 \longrightarrow \text{? mol } KClO_3 \longrightarrow \text{? g } KClO_3$$

$$1.91 \times 10^{-2} \text{ mol } O_2 \times \frac{2 \text{ mol } KClO_3}{3 \text{ mol } O_2} \times \frac{122.6\text{g } KClO_3}{1 \text{ mol } KClO_3} = 1.58\text{g } KClO_3$$

---

## ANSWERS TO SELF-TEST

1. c
2. c
3. d
4. c
5. a
6. c
7. d
8. d
9. b
10. a

# LIQUIDS AND SOLIDS

# CHAPTER 11



## 11.1 Intermolecular Forces

Attractive forces that exist between molecules are called intermolecular forces.

Attractions between oppositely charged ends of dipoles are called dipolar attractions. Dipolar attractions are weaker than attractions between ions and operate over shorter distances than ionic attractions.

When molecules of similar molar mass are compared, the strength of dipolar attractions increases as the polarity of the molecule increases. (Molecular polarity depends on the polarity of the bonds in the molecule and the shape of the molecule.) The stronger the intermolecular attractions, the higher the boiling point (Table 11-1).

Hydrogen bonds are especially strong dipolar attractions occurring when hydrogen bonds to fluorine, oxygen, or nitrogen (Table 11-2).

London dispersion forces are forces resulting from attractions between instantaneous dipoles. These instantaneous dipoles occur when electrons in adjacent molecules repel each other causing shifts in electron density.

Intermolecular attractions due to London dispersion forces increase with molecular mass and size (Table 11-3).

In order of increasing intermolecular forces,

London forces < dipolar forces < H-bonding

## 11.2    General Classes of Liquids

Nonpolar liquids have molecules attracted to each other by London dispersion forces.  These liquids have relatively low boiling points.

Polar liquids have molecules attracted to each other by London forces and dipolar forces.  These liquids have relatively high boiling points.  Polar liquids which have hydrogen bonding have the highest boiling points.

## 11.3    Densities of Liquids

Liquids have small regions of fairly ordered molecules and other regions in which the molecules are completely disordered.  The molecules in the disordered regions are loosely packed.  Therefore, most liquids are less dense than solids.

Liquid water has an unusually high density.  Hydrogen bonding in water enables the molecules to pack more closely than in most liquids.

## 11.4    Surface Tension

Surface tension is caused by an imbalance of intermolecular forces at the surface of a liquid.  Molecules on the surface are being attracted by molecules only on the sides and underneath (Figure 11-8).  This results in a net inward pull on the surface molecules.

## 11.5    Viscosity

The resistance to flow exerted by a fluid is called viscosity.

The higher the viscosity, the more slowly molecules move past each other.

Large, unsymmetrical molecules have high viscosities (Table 11-4).

Viscosity increases as temperature decreases.

## 11.6    Vaporization

Vaporization is the process in which molecules pass from the liquid to the gaseous state.

In an open container, liquid molecules with enough kinetic energy can break away from the surface of the liquid.  The vapor molecules disperse throughout the atmosphere.

In a closed container, the liquid and vapor phase establish a dynamic equilibrium (Figure 11-10).

The pressure exerted by a vapor in equilibrium with its liquid is called the vapor pressure of the liquid.

The vapor pressure of a liquid depends on the strength of the intermolecular forces in the liquid.  Substances with relatively weak intermolecular forces have high vapor pressures, low boiling points, and evaporate easily (Table 11-5).

The vapor pressure of a substance increases with temperature (Table 11-5 and Figure 11-11).

## 11.7  Boiling Point

The boiling point of a liquid is the temperature at which the vapor pressure of the liquid is the same as the pressure of the atmosphere.

The boiling point decreases as atmospheric pressure decreases, and vice versa.

## 11.8  Heat of Vaporization

When a liquid is heated at its boiling point in an open container the temperature remains constant until all of the liquid is changed into vapor.  The added heat is used to overcome the intermolecular forces among the liquid molecules, allowing them to become vapor molecules.

The molar heat of vaporization, $\Delta H_{vap}$, is the amount of heat required to vaporize one mole of a boiling liquid under one atmosphere of pressure (Table 11-7).

The larger the value of $\Delta H_{vap}$, the greater the intermolecular forces among the liquid molecules.

## 11.9  The Solid State

Five types of crystalline solids exist:
1. Ionic crystals have ions held at lattice points by electrical forces of attraction between cations and anions.
2. Molecular crystals have molecules held at lattice points by intermolecular forces (Figure 11-14).
3. Covalent crystals have atoms held at lattice points by covalent bonds (Figure 11-15).
4. Metallic crystals have metal cations at lattice points surrounded by a "sea of electrons" (Figure 11-16).
5. Atomic crystals have single nonmetal atoms held at lattice points by London dispersion forces.

Amorphous solids do not have an orderly arrangement of particles and thus do not exist as crystals.

11.10   The Melting Process

When heat is added to a crystalline solid, the particles
within the solid can acquire enough energy to overcome the
crystal lattice forces.  This process, called melting,
changes the solid into a liquid.

11.11   Heat of Fusion

The melting point of a substance is the temperature at which
liquid and solid are in equilibrium.

When heat is added to a solid at its melting point, the
temperature remains constant until all of the solid is
changed into liquid.  The added heat is used to overcome the
forces among the particles in the solid state.

The molar heat of fusion, $\Delta H_{fus}$, is the amount of heat needed
to melt one mole of a crystalline solid at its melting point
(Table 11-8).

The molar heat of vaporization of a substance is usually
larger than the molar heat of fusion.

11.12   Melting Point

Crystalline solids melt over a very narrow temperature range.
The crystal lattice forces within a crystal are approximately
equal in force, so they are disrupted at very nearly the same
temperature.

The higher the melting point, the stronger the crystal lattice
forces.

An amorphous solid is held together by different forces with
different strengths, so melting occurs over a broad temperature range.

11.13   Calculations Based on Phase Changes (Optional)

When a pure substance is heated, the temperature increases
according to its specific heat until a phase change occurs.

A heating curve shows temperature changes as a function of
time when heat is added to a substance at a constant rate
(Figure 11-19).

---

SOLUTIONS TO SELECTED STUDY QUESTIONS AND PROBLEMS
and
PRACTICE PROBLEMS

---

3.      Predict the type(s) of intermolecular forces that would be
        exerted by each of the following molecules for others of the
        same kind.

        To predict intermolecular forces, classify the molecule as
        polar or nonpolar based on bond type and molecular geometry

(see Chapter 6).  From this classification, the following forces can be predicted:

| molecule | intermolecular forces |
|---|---|
| nonpolar | London dispersion forces |
| polar without H-N, H-O, or H-F bonds | London dispersion forces, dipolar attractions |
| polar with H-N, H-O, or H-F bonds | London dispersion forces, hydrogen bonds |

a.  $CCl_4$

CCl$_4$ is a tetrahedral molecule with polar C-Cl bonds. The chlorine atoms withdraw bonding electrons with equal force in opposite directions, resulting in a nonpolar molecule.  One CCl$_4$ molecule attracts another by London dispersion forces only.

b.  $CH_3$-OH

The electron dot formula for $CH_3$-OH is

$$
\begin{array}{c}
H \\
| \\
H-C-\ddot{O}-H \\
| \\
H
\end{array}
$$

The polar bonds and geometry of CH$_3$OH result in a polar molecule.  In addition, hydrogen is found bonded to oxygen in this molecule.

The intermolecular forces of attraction are London dispersion forces and hydrogen bonds.  Notice in the electron dot formulas below, that only the hydrogen atoms bonded to oxygen are capable of hydrogen bonding.

$$
\begin{array}{ccc}
:\ddot{O}-H & ----- & :\ddot{O}-H \\
| & & | \\
H-C-H & & H-C-H \\
| & & | \\
H & & H
\end{array}
$$

Consult Appendix C in the text for answers to parts c, d, e, and f.

19.  Why does water have a lower vapor pressure than ethyl alcohol or ethyl ether at a given temperature?

The vapor pressure of a liquid depends on the strength of the intermolecular forces in the liquid.  Compare the electron dot formulas below

$$
\begin{array}{lll}
H-\ddot{O}: & H-\ddot{O}: & H-\overset{\displaystyle H}{\underset{\displaystyle H}{C}}-\ddot{O}: \\
| & | & \\
H & H-C-H & \\
& | & H-C-H \\
\text{water} & H & | \\
& & H \\
& \text{ethyl alcohol} & \text{ethyl ether}
\end{array}
$$

Water has two hydrogen atoms capable of forming hydrogen bonds. Ethyl alcohol has one hydrogen atom capable of forming hydrogen bonds. Ethyl ether has no hydrogen atoms capable of forming hydrogen bonds. Therefore, the intermolecular forces are strongest in water and weakest in ethyl ether. Molecules with strong intermolecular forces have low vapor pressures.

***Practice Problem A

Which of the following compounds has the higher boiling point? Explain

a.  Ar or Xe

b.  HCl or $Cl_2$

c.  HCl or HF

31.  Calculate the amount of heat in kilojoules needed to vaporize 3.75g of $H_2O$ at 100.0°C.

To calculate the amount of heat needed to vaporize 3.75g of $H_2O$ at its boiling point, consult Table 11-7. The heat of vaporization, $\Delta H_{vap}$, of water is 40.6 kJ/mol. The quantity of water, given in grams, must be converted to moles.

$$3.75g\ \cancel{H_2O} \times \frac{1\ mol\ H_2O}{18.0g\ \cancel{H_2O}} = 0.208\ mol\ H_2O$$

The heat of vaporization relates heat in kJ to molar quantities. Think of $\dfrac{40.6 \text{ kJ}}{1 \text{ mol}}$ as a conversion factor.

$$0.208 \text{ mol } H_2O \times \frac{? \text{ kJ}}{? \text{ mol } H_2O} =$$

$$0.208 \text{ mol } H_2O \times \frac{40.6 \text{ kJ}}{1 \text{ mol } H_2O} = 8.44 \text{ kJ}$$

Or, in a multi-step conversion

$$3.75g \text{ } H_2O \times \frac{1 \text{ mol } H_2O}{18.0g \text{ } H_2O} \times \frac{40.6 \text{ kJ}}{1 \text{ mol } H_2O} = 8.46 \text{ kJ}$$

51. Calculate the amount of heat in kilojoules needed to melt 50.0g of ice at 0.0°C.

The amount of heat to melt a compound at its melting point depends on the heat of fusion, $\Delta H_{fus}$. Consult Table 11-8. The heat of fusion for water is 5.98 kJ/mol.

$$50.0g \text{ } H_2O \times \frac{1 \text{ mol } H_2O}{18.0g \text{ } H_2O} \times \frac{5.98 \text{ kJ}}{1 \text{ mol } H_2O} = 16.6 \text{ kJ}$$

Notice the similarities in the solutions for #31 and #51.

***Practice Problem B

Calculate the amount of heat in kilojoules needed to melt 10.0g of ethyl alcohol at its melting point. Then calculate the amount of heat in kilojoules needed to vaporize 10.0g of ethyl alcohol at its boiling point. Which process requires the greater amount of heat? Why? (Data needed for solution is given below.)

| Formula | Formula Weight | Melting Point (°C) | Boiling Point (°C) | $\Delta H_{fus}$ (kJ/mol) | $\Delta H_{vap}$ (kJ/mol) |
|---|---|---|---|---|---|
| $CH_3CH_2OH$ | 46.0 | −117 | 78.5 | 5.02 | 40.5 |

59. Use the following data for oxygen to calculate the heat (kJ) liberated when 100.0g of $O_2$ are cooled from 25.0°C to -200.0°C.

| | |
|---|---|
| melting point | -219°C |
| boiling point | -183°C |
| $\Delta H_{vap}$ | 6.82 kJ/mol |
| $\Delta H_{fus}$ | 0.444 kJ/mol |
| specific heat of $O_2$ (gas) | 0.228 $\dfrac{cal}{g \cdot °C}$ |
| specific heat of $O_2$ (liquid) | 0.35 $\dfrac{cal}{g \cdot °C}$ |

A number line labeled with the melting point, boiling point, initial temperature and final temperature graphically outlines the problem.

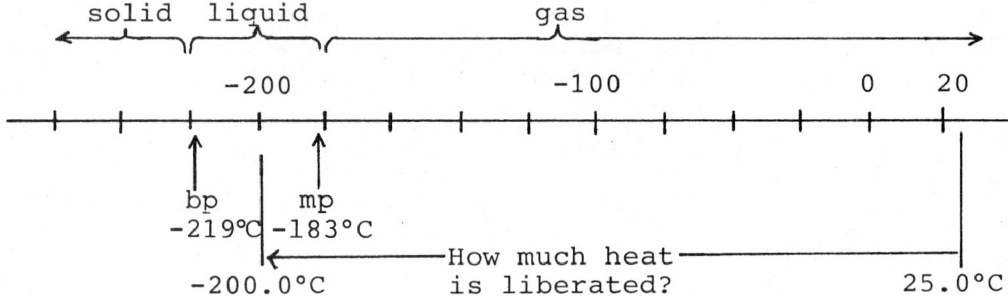


Consider the cooling process in several steps.

1. Cooling the gas from 25.0°C to -183°C (boiling point).
2. Gas to the liquid state change (while the temperature remains at -183°C).
3. Cooling the liquid from -183°C to -200.0°C.

(Further cooling would require additional calculations to change the state from liquid to solid and cool the solid.)

Calculate the amount of heat liberated in each of the steps and then add the amounts together.

1. $gas_{25.0°C} \longrightarrow gas_{-183°C}$

   Heat liberated = mass x specific heat x $\Delta T$

   $$= (100.0g)\left(0.228 \, \frac{cal}{g \cdot °C}\right)(208°C)$$

   $$= 4742 \text{ cal} = 4.74 \times 10^3 \text{ cal}$$

   Convert $4.74 \times 10^3$ cal to kJ.

   $$4.74 \times 10^3 \text{ cal} \times \frac{4.185 \text{ J}}{1 \text{ cal}} \times \frac{1 \text{ kJ}}{10^3 \text{ J}} = 19.8 \text{ kJ}$$

2. $gas_{-183°C} \longrightarrow liquid_{-183°C}$

   Heat liberated = $100.0g \, O_2 \times \dfrac{1 \text{ mol } O_2}{32.0g} \times \dfrac{6.82 \text{ KJ}}{mol} = 21.3 \text{ KJ}$

3. $\text{liquid}_{-183°C} \longrightarrow \text{liquid}_{-200.0°C}$

Heat liberated = mass x specific heat x $\Delta$T

$$= (100.0g) \left(0.35 \frac{cal}{g \cdot °C}\right)(17°C)$$

$$= 595 \text{ cal}$$

Convert 595 cal to kJ

$$595 \text{ cal} \times \frac{4.185 \text{ J}}{1 \text{ cal}} \times \frac{1 \text{ kJ}}{10^3 \text{ J}} = 2.49 \text{ kJ}$$

The total amount of heat liberated:

| | |
|---|---|
| step 1 | 19.8 kJ |
| step 2 | 21.3 kJ |
| step 3 | 2.49 kJ |
| | 43.59 kJ = 43.6 kJ |

***Practice Problem C

Use the following data for water to calculate the heat (kJ) required when 55.0g of $H_2O$ are heated from -20.0°C to 120°C.

| | |
|---|---|
| $\Delta H_{fus}$ | 5.98 kJ/mol |
| $\Delta H_{vap}$ | 40.6 kJ/mol |
| specific heat of $H_2O$ (solid) | 0.492 $\frac{cal}{g \cdot °C}$ |
| specific heat of $H_2O$ (liquid) | 1.00 $\frac{cal}{g \cdot °C}$ |
| specific heat of $H_2O$ (gas) | 0.485 $\frac{cal}{g \cdot °C}$ |

---

SELF-TEST

---

Circle the correct answer in the following multiple choice questions.

1.  Which of the following substances would have the smallest molar heat of vaporization?

    a.  $H_2O$
    b.  $NH_3$
    c.  $CH_4$
    d.  HF

2.  Which of the following substances would be expected to have the highest boiling point?

    a.  $I_2$
    b.  $CH_3OH$
    c.  $CH_3Cl$
    d.  $NH_3$

3.  The molar enthalpy of vaporization of mercury is 59.1 kJ/mol. Calculate the amount of heat required to vaporize 100.0g of mercury at its boiling point.

    a.  29.5 kJ
    b.  118 kJ
    c.  5910 kJ
    d.  0.295 kJ

4.  Which of the following substances has only London dispersion forces present?

    a.  $F_2O$
    b.  HF
    c.  NaCl
    d.  $Br_2$

5.  Which of the following statements is true for viscosity?

    a.  Viscosity increases as temperature increases.
    b.  The higher the viscosity, the more slowly molecules move past each other.
    c.  Symmetrical molecules have high viscosities.
    d.  Ether, $CH_3OCH_3$, should be more viscous than ethyl alcohol, $CH_3OH$.

6.  The vapor pressure of a liquid

    a.  can only be measured in an open container.
    b.  decreases as temperature increases.
    c.  is high for substances with weak intermolecular forces.
    d.  is low for liquids with low boiling points.

7.      Which of the following compounds would be expected to have
        the highest melting point?

        a.  ionic compound
        b.  nonpolar molecular compound
        c.  polar molecular compound
        d.  atomic solid

8.      A heating curve for mercury has two horizontal regions; one
        at 234 K, the other at 630 K. The horizontal region at 234 K
        corresponds to

        a.  heating $Hg_{(l)}$

        b.  heating $Hg_{(g)}$

        c.  $Hg_{(s)} \longrightarrow Hg_{(l)}$

        d.  $Hg_{(l)} \longrightarrow Hg_{(g)}$

9.      Calculate the amount of heat required when 150.0g of ice is
        heated from -25.0°C to 0°C.  Use data in Table 2-5.

        a.  $1.85 \times 10^3$ kJ
        b.  7.72 kJ
        c.  3.75 kJ
        d.  1.85 kJ

10.     Calculate the amount of heat required when 50.0g of $H_2O_{(s)}$ is
        heated from -15.0°C to 60.0°C.  Use data in Tables
        11-7, 11-8, and 2-5.

        a.  14.1 kJ
        b.  28.0 kJ
        c.  19.4 kJ
        d.  30.7 kJ

---

## ANSWERS TO PRACTICE PROBLEMS

Practice Problem A

        a.  Xe

           The intermolecular forces of attraction in both atoms are
           London dispersion forces.  Since Xe atoms are larger than
           Ar atoms the electron distribution fluctuates to greater
           extremes in Xe, making the London forces stronger.  The
           stronger the forces of attraction between particles, the
           higher the boiling point.

        b.  HCl

           HCl is a polar molecule with London dispersion forces and
           dipolar attractions.  $Cl_2$ is a nonpolar molecule with
           only London dispersion forces. London dispersion forces
           are weaker than dipolar attractions, so the boiling point
           in HCl is higher.

c.   HF

HF has hydrogen bonding in addition to London dispersion forces.  HCl has dipolar attractions and London dispersion forces.  Since hydrogen bonds are stronger attractive forces than dipolar attractions, HF has a higher boiling point.

## Practice Problem B

Heat required to melt 10.0g of ethyl alcohol:

$$10.0g\ CH_3CH_2OH \times 1\ mol\ \frac{CH_3CH_2OH}{46.0g} \times \frac{5.02\ kJ}{1\ mol} = 1.09\ kJ$$

Heat required to vaporize 10.0g of ethyl alcohol:

$$10.0g\ CH_3CH_2OH \times 1\ mol\ \frac{CH_3CH_2OH}{46.0g} \times \frac{40.5\ kJ}{1\ mol} = 8.80\ kJ$$

Vaporization of ethyl alcohol requires more heat than melting.  In melting, molecules are moved apart somewhat when the crystal lattice is disrupted, but the molecules are still in contact with each other.  In vaporization, the attractive forces between the particles must be completely overcome.  Compare the values of $\Delta H_{fus}$ to $\Delta H_{vap}$.  More heat must be applied to vaporize ethyl alcohol.

## Practice Problem C

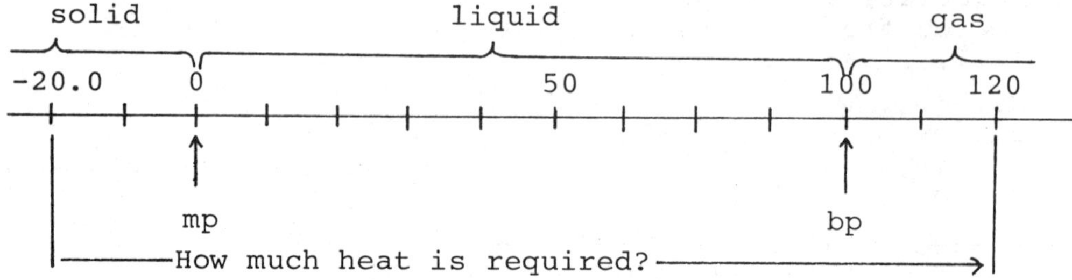


1.   $solid_{-20.0°C} \longrightarrow solid_{0°C}$

Heat required $= (55.0g\ H_2O)\left(\dfrac{0.492\ cal}{g \cdot °C}\right)(20.0°C) = 541\ cal$

Convert 541 cal to kJ:

$$541\ cal \times \frac{4.185\ J}{1\ cal} \times \frac{1\ kJ}{10^3\ J} = 2.26\ kJ$$

2.   $solid_{0°C} \longrightarrow liquid_{0°C}$

Heat required $= 55.0g\ H_2O \times \dfrac{1\ mol\ H_2O}{18.0g} \times \dfrac{5.98\ kJ}{1\ mol} = 18.3\ kJ$

3.   $liquid_{0°C} \longrightarrow liquid_{100°C}$

Heat required $= (55.0g\ H_2O)\left(1.00\ \dfrac{cal}{g \cdot °C}\right)(100°C) = 5500\ cal$

Convert 5500 cal to kJ:

$$5500 \text{ cal} \times \frac{4.185 \text{ J}}{1 \text{ cal}} \times \frac{1 \text{ kJ}}{10^3 \text{ J}} = 23.0 \text{ kJ}$$

4. $\text{liquid}_{100°C} \longrightarrow \text{gas}_{100°}$

Heat required = $55.0 \text{g } H_2O \times \dfrac{1 \text{ mol } H_2O}{18.0 \text{g}} \times \dfrac{40.6 \text{ kJ}}{1 \text{ mol}} = 124 \text{ kJ}$

5. $\text{gas}_{100°C} \longrightarrow \text{gas}_{120°C}$

Heat required = $(55.0 \text{g}) \left( 0.485 \dfrac{\text{cal}}{\text{g}\cdot°C} \right) (20°C) = 534 \text{ cal}$

Convert 534 cal to kJ:

$$534 \text{ cal} \times \frac{4.185 \text{ J}}{1 \text{ cal}} \times \frac{1 \text{ kJ}}{10^3 \text{ J}} = 2.24 \text{ kJ}$$

The total amount of heat required:

|        |        |    |
|--------|--------|----|
| step 1 | 2.26   | kJ |
| step 2 | 18.3   | kJ |
| step 3 | 23.0   | kJ |
| step 4 | 124    | kJ |
| step 5 | 2.24   | kJ |
|        | 169.80 | kJ = 170 kJ |

The answer is rounded to 170 kJ.  In addition, the answer can have no more decimal places than the answer with the least number of decimal places.

---

## ANSWERS TO SELF-TEST

1. c
2. b
3. a
4. d
5. b
6. c
7. a
8. c
9. b
10. d

# WATER AND AQUEOUS SOLUTIONS

# CHAPTER
# 12

---

## SUMMARY OUTLINE

### 12.1   Water:  Structure and Properties

Liquid water consists of aggregates of hydrogen-bonded molecules with individual molecules interspersed among the aggregates and in equilibrium with them. An average of three hydrogen bonds are formed by each molecule in liquid water (Figure 12-3).

Solid water (ice) is more ordered than liquid water. Each molecule forms four hydrogen bonds, resulting in larger spaces between the molecules in the solid state than in the liquid state (Figure 12-3).

Ice is less dense than liquid water.

The hydrogen bonding in water accounts for the relatively high melting point, boiling point, $\Delta H_{fus}$, and $\Delta H_{vap}$. The geometry of water molecules along with hydrogen bonding enable water molecules to pack more tightly than molecules of most liquids. Water has a higher density than expected based on its formula weight (Table 12-1).

### 12.2   Properties of Solutions

Solutions are homogeneous mixtures.

A solvent is the major component of a mixture and a solute is the minor component (Table 12-2).

A solution in which water is the solvent is called an aqueous solution.

Solutions can be colorless or colored, but all solutions are clear.

Dissolved particles in solution can be ions, atoms, or molecules.

12.3    Solution Formation

A solution will form if the attractive forces between solute particles and solvent molecules are stronger than those among solute particles and those among solvent molecules.

Polar water molecules are capable of dissolving polar covalent and ionic substances.  Ions are attracted by the partial charges present in the polar water molecule.  Polar solutes are attracted by water through dipolar attractions or in some cases, hydrogen bonds (Figures 12-5, 12-6, and 12-7).

The process in which solvent molecules surround solute particles in solution is called solvation.  When the solvent is water, the term hydration is used.

The rate at which a solute dissolves in a liquid solvent depends on the surface area of the undissolved solute, the amount of agitation, and the temperature.

12.4    Solubility

The solubility of a solute is the maximum amount of solute that will dissolve in a given amount of solvent at a particular temperature (Table 12-4).

Substances having water solubilities of 1g solute or greater per 100 mL of water are described as soluble.  Substances with solubilities less than that are described as insoluble.

Miscible liquids are two liquids that are mutually soluble.

When a solute is added to a solvent in amounts in excess of the solubility, a dynamic equilibrium is established between the solution and the undissolved solute.  The solution is described as a saturated solution (Figure 12-8).

The heat change that occurs when one mole of a solute dissolves in a solvent is called the molar heat of solution, $\Delta H_{solution}$, of the solute.  If heat is released during dissolving, the solution becomes warm; the process is exothermic, and $\Delta H_{solution}$ is negative.  If heat is absorbed, the solution becomes cool; the process is endothermic, and $\Delta H_{solution}$ is positive (Table 12-5).  Generally, the more heat released during dissolving, the greater the solubility.

## 12.5 Factors Affecting Solubility

Solution formation causes a volume change when gases dissolve in liquids or solids. For these solutions, an increase in external pressure will increase the solubility of the gas (Figure 12-9).

The solubility of a solution is affected by temperature increases in the following ways:

1. For solutes with $+\Delta H_{solution}$, solubility increases.
2. For solutes with $-\Delta H_{solution}$, solubility decreases.

Hydrophilic substances are soluble in water, hydrophobic substances are not. In general, polar solvents are likely to dissolve polar and ionic solutes; nonpolar solvents dissolve only nonpolar solutes.

## 12.6 Concentrations of Solutions: Molarity

The concentration of a solution is the amount of solute in a given quantity of solution.

Molarity, M, is the number of moles of solute per liter of solution.

$$M = \frac{mol\ solute}{volume\ solution}$$

$$M = \frac{n}{V}$$

where n = moles of solute and V = volume of solution. M has units of mol/L.

Molarity calculations involve three variables: molarity, moles of solute, and volume of solution. Two types of calculations result:

1. Calculating M; given a mass (or number of moles) of solute, and a volume of solution.
2. Given M, and one other variable; calculating the third variable.

To solve the second type, write molarity as a ratio. Use this ratio like a conversion factor.

## 12.7 Concentrations of Solutions: Percent Concentration

The weight-to-volume percent concentration, %(w/v), is defined as

$$\%(w/v) = \frac{g\ solute}{mL\ solution} \times 100\%$$

The volume-to-volume percent concentration, %(v/v), is defined as

$$\%(v/v) = \frac{mL\ solute}{mL\ solution} \times 100\%$$

The weight-to-weight percent concentration, %(w/w), is defined as

$$\%(w/w) = \frac{g\ solute}{g\ solution} \times 100\%$$

12.8    Dilutions

Dilutions are accomplished by adding solvent to concentrated stock solution, thus lowering the concentration of the stock solution.

When water is added to a concentrated solution, the total amount of dissolved solute remains the same, it is just distributed throughout a larger volume of solution.  Thus

$$n_c = n_d,$$

where c indicates concentrated and d indicates dilute.  The following relationship can be used to solve dilution problems,

$$M_c V_c = M_d V_d \text{ or}$$
$$C_c V_c = C_d V_d$$

where C represents any concentration expression.

12.9    Stoichiometric Calculations Involving Solutions

Solution stoichiometry can be summarized by the following "mole map".

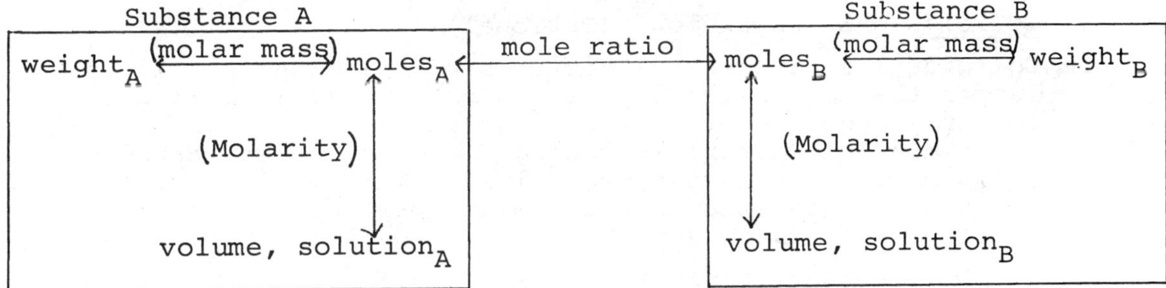


12.10    Colligative Properties of Solutions (Optional)

Properties of solutions that depend only on the number of solute particles and not their identity are called colligative properties.  Colligative properties include vapor-pressure lowering, boiling-point elevation, freezing-point lowering and osmotic pressure.

A nonvolatile solute present in a liquid solution reduces the vapor pressure of the solution.  This reduced vapor pressure results in solutions with higher boiling points and lower freezing points than those of the pure solvent.

Freezing point depressions can be used to determine the molar mass of the nonvolatile solute.  For a molecular solute, the amount of freezing-point lowering is

$$\Delta t_f = K_f m$$

$\Delta t_f$ = amount by which freezing point is lowered in °C.

$K_f$ = freezing point depression constant for the solvent in units of $\dfrac{°C \cdot Kg\ solvent}{mol\ solute}$ (Table 12-6).

m = molality of the solute.  Molality, a concentration expression, is defined as moles of solute per kilogram of solvent.

$$M = \frac{\text{mol solute}}{\text{Kg solvent}}$$

$$\Delta t_f = K_f \left(\frac{\text{mol solute}}{\text{kg solvent}}\right)$$

$$= K_f \,(\text{mol solute}) \left(\frac{1}{\text{kg solvent}}\right)$$

$$\Delta t_f = K_f \left(\frac{\text{g solute}}{\text{molar mass solute}}\right) \left(\frac{1}{\text{kg solvent}}\right)$$

To find molar mass of solute, the $\Delta t_f$ expression can be rearranged.

$$\text{molar mass solute} = K_f \left(\frac{\text{g solute}}{\Delta t_f}\right) \left(\frac{1}{\text{kg solvent}}\right)$$

Osmotic pressure is the minimum amount of pressure (applied to the surface of a solution) needed to prevent passage of water across an osmotic membrane into a solution.

## 12.11  <u>Colloids and Suspensions</u> (Optional)

Colloids are homogeneous mixtures of dispersed particles larger than most molecules (1-100nm).  Colloidal dispersions appear cloudy and are not true solutions.  Colloidal particles are bombarded by rapidly moving fluid molecules, causing them to move randomly through the fluid.  This motion, called Brownian motion, prevents colloidal particles from settling (Table 12-7).

Light passing through true solutions is not visible from the side because the solute particles are not large enough to reflect light.  Colloidal particles can reflect light, so when light passes through colloidal dispersions, the light beam is visible.  This is called the Tyndall effect.

When particles larger than colloidal particles are scattered through a fluid, the heterogeneous mixture is called a suspension.  The suspended particles are visible and will settle out.

---

SOLUTIONS TO SELECTED STUDY QUESTIONS AND PROBLEMS
and
PRACTICE PROBLEMS

---

25.  Calculate the molarity of each of the following solutions.

c.  15.0g of LiBr in 500.0 mL of solution

To calculate molarity,

$$M = \frac{\text{mol solute}}{\text{volume of solution (L)}}$$

first calculate mol of LiBr.

$$15.0 \text{g LiBr} \times \frac{1 \text{ mol LiBr}}{86.8 \text{g}} = 0.173 \text{ mol LiBr}$$

Since molarity has units of mol/L, the volume of the solution must be converted to liters.

$$500.0 \text{ mL} \times \frac{10^{-3} \text{ L}}{1 \text{ mL}} = 0.5000 \text{ L}$$

The molarity can now be calculated.

$$M = \frac{0.173 \text{ mol LiBr}}{0.5000 \text{ L solution}}$$

$$= \frac{0.346 \text{ mol LiBr}}{1 \text{ L solution}} = 0.346 \text{ M LiBr}$$

Notice that 0.344 M LiBr (read zero point three four four molar lithium bromide) is an abbreviated way of writing

$$\frac{0.344 \text{ mol LiBr}}{1 \text{ L LiBr solution}}$$

The longer method is seldom written when referring to the concentration of a particular solution.  The ratio is used when solving problems where molarity is given and moles of solute or volume of solution is required (see problem 26).

Consult Appendix C in the text for answers to parts a, b, d, and e.

26.  Calculate the number of grams of solute needed to prepare each of the following solutions.

a.  1.00 L of 0.500 M glucose, $C_6H_{12}O_6$

Begin by writing the molarity as a conversion factor.

$$0.500 \text{ M } C_6H_{12}O_6 = \frac{0.500 \text{ mol } C_6H_{12}O_6}{1 \text{ L solution}}$$

This conversion factor can be used to interconvert moles (or mass) of $C_6H_{12}O_6$ to volume of glucose solution.

To solve for grams of glucose, multiply the given data, (in this case, 1.00 L solution) by the conversion factor

$$1.00 \text{ L solution} \times \frac{0.500 \text{ mol } C_6H_{12}O_6}{1 \text{ L solution}}$$

Since grams of $C_6H_{12}O_6$ are required, include one additional conversion factor to complete the calculation.

$$1.00 \text{ L solution} \times \frac{0.500 \text{ mol } C_6H_{12}O_6}{1 \text{ L solution}} \times \frac{180 \text{g } C_6H_{12}O_6}{1 \text{ mol } C_6H_{12}O_6} =$$

$$90.0 \text{g } C_6H_{12}O_6$$

The solution to this problem can be summarized as

$$\text{L solution} \xrightarrow{(M)} \text{mol } C_6H_{12}O_6 \xrightarrow{(\text{molar mass})} \text{g } C_6H_{12}O_6$$

b.   800.0 mL of 0.100 M NaCl

Given information:  800.0 mL NaCl solution
Conversion factor:   $\dfrac{0.100 \text{ mol NaCl}}{1 \text{ L solution}}$

The given volume must first be converted to L, then the problem is solved as in part a above.

$$\text{mL solution} \longrightarrow \text{L solution} \xrightarrow{(M)} \text{mol NaCl} \longrightarrow \text{g NaCl}$$

$$800.0 \text{ mL} \times \frac{10^{-3} \text{ L}}{1 \text{ mL}} \times \frac{0.100 \text{ mol NaCl}}{1 \text{ L solution}} \times \frac{58.44 \text{g NaCl}}{1 \text{ mol NaCl}} = 4.68 \text{g NaCl}$$

Parts c, d, e and f are solved in a similar manner.

27.   Calculate the number of moles of solute in each of the following solutions.

a.   150.0 mL of 0.100 M NaCl

Given information:  150.0 mL NaCl solution
Conversion factor:   $\dfrac{0.100 \text{ mol NaCl}}{1 \text{ L solution}}$

$$\text{mL solution} \longrightarrow \text{L solution} \xrightarrow{(M)} \text{mol NaCl}$$

$$150.0 \text{ mL} \times \frac{10^{-3} \text{ L}}{1 \text{ mL}} \times \frac{0.100 \text{ mol NaCl}}{1 \text{ L solution}} = 1.50 \times 10^{-2} \text{ mol NaCl}$$

In this calculation, the conversion from mol NaCl $\longrightarrow$ g NaCl was omitted because only moles was required in the problem.

Consult Appendix C in the text for answers to parts b, c, d, e and f.

***Practice Problem A

a.   Calculate the molarity of a NaOH solution prepared by dissolving 25.0g of NaOH in enough water to give 5.00 L of solution.

b.  How many grams of NaOH are present in 200.0 mL of a
1.25 M NaOH solution?

c.  How many liters of 1.25 M NaOH can be prepared from
50.0g of NaOH?

28.   Calculate the %(w/v) concentration of each of the following
solutions.

a.  17.0g of solute in 150.0 mL of solution.

By definition, %(w/v) is

$$\%(w/v) = \frac{g\ solute}{mL\ solution} \times 100$$

$$= \frac{17.0g\ solute}{150.0\ mL\ solution} \times 100$$

$$= 11.3\%(w/v)$$

Parts b, c, d, e and f are solved in the same way.

29.   Hydrogen peroxide solution for bleaching hair is usually at a
concentration of 5%(w/v) in water.  How many grams of hydrogen
peroxide, $H_2O_2$, are present in 80.0 mL of this solution?

Like molarity, percent concentrations can be used as conversion factors.

$$5\%(w/v) = \frac{5g\ H_2O_2}{100\ mL\ solution}$$

Given information:   80.0 mL solution
Conversion factor:   $\dfrac{5g\ H_2O_2}{100\ mL\ solution}$

mL solution $\xrightarrow{\%(w/v)}$ g $H_2O_2$

80.0 mL solution x $\dfrac{5g\ H_2O_2}{100\ mL\ solution}$ = 4.00g $H_2O_2$

***Practice Problem B

a.  Calculate the %(w/v) for a sodium chloride solution used
    as an intraveneous solution for certain illnesses.
    250.0 mL of the solution contains 2.30g of NaCl.

b.  How many grams of a 0.92%(w/v) NaCl solution are required
    to prepare 425 mL of this solution?

c.  How many mL of a 0.92%(w/v) NaCl solution can be prepared
    from 5.00g NaCl?

31.   Calculate the %(v/v) concentration of each of the following solutions.

    a.   17.0 mL of solute in 150.0 mL of solution.

       By definition, %(v/v) is

$$\%(v/v) = \frac{mL\ solute}{mL\ solution} \times 100$$

$$= \frac{17.0\ mL\ solute}{150.0\ mL\ solution} \times 100$$

$$= 11.3\%(v/v)$$

Consult Appendix C in the text for answers to parts b, c, d, e and f.

32.   Calculate the %(w/w) concentration of each of the following solutions.

    a.   17.0g of solute in 150.0g of solution.

       By definition, %(w/w) is

$$\%(w/w) = \frac{g\ solute}{g\ solution} \times 100$$

$$= \frac{17.0g\ solute}{150.0g\ solution} \times 100$$

$$= 11.3\%(w/w)$$

Parts b, c, d, e and f are solved in the same way.

The definitions for %(w/v), %(v/v) and %(w/w) can be easily distinguished and recalled by the letters in parenthesis.  In each case the first letter refers to solute (written in the numerator) and the second letter refers to solution (written in the denominator).  The letter w always refers to mass in grams, and the letter v always refers to volume in mL.  For example a new concentration expression, %(v/w), could be defined as

$$\%(v/w) = \frac{mL\ solute}{g\ solution} \times 100$$

***Practice Problem C

A miscible solution of ethyl alcohol ($CH_3CH_2OH$) and water was prepared by dissolving 50.0 mL of ethyl alcohol (density = 0.789g/mL) in water.  The resulting solution had a volume of 125.0 mL and a density of 0.950g/mL.

    a.   Calculate the %(v/v) concentration of the solution.

b.  Calculate the %(w/v) concentration of the solution.

c.  Calculate the %(w/w) concentration of the solution.

d.  What mass of ethyl alcohol would be required to prepare 200.0 mL of this solution?  (Hint:  Use a calculated percent concentration expression from above as a conversion factor.)

34.  Calculate the final volumes for diluting each of the solutions in question 27 to 0.0500 M.

a.  150.0 mL of 0.100 M NaCl

$$C_c V_c = C_d V_d$$

State the problem.

$C_c$ = 0.100 M NaCl               $C_d$ = 0.0500 M NaCl

$V_c$ = 150.0 mL solution          $V_d$ = ?

Rearrange the equation above, solving for $V_d$.

$$V_d = V_c \left( \frac{C_c}{C_d} \right)$$

$$= 150.0 \text{ mL} \left( \frac{0.100 \text{ M}}{0.0500 \text{ M}} \right)$$

$$= 300 \text{ mL} = 3.00 \times 10^2 \text{ mL}$$

b.  200.0 ml of 0.250 M $KNO_3$

State the problem.

$C_c$ = 0.250 M $KNO_3$           $C_d$ = 0.0500M
$V_c$ = 200.0 ml                 $V_d$ = ?

$$V_c = V_c\left(\frac{C_c}{C_d}\right)$$

$$= 200.0 \text{ ml}\left(\frac{0.250 \cancel{M}}{0.0500 \cancel{M}}\right)$$

$$= 1.00 \times 10^3 \text{ mL}$$

Consult Appendix C in the text for answers to parts c, d, e and f.

***Practice Problem D

a.  What will be the concentration of a solution prepared by diluting 100.0 mL of a 7.50% (w/v) solution to 250.0 mL?

b.  To what volume would 75.0 mL of a 15.0% (v/v) solution be diluted to prepare a 10.0% (v/v) solution?

38.  What volume of 0.250 M KCl is required to react completely with 165 mL of 0.300 M $Pb(NO_3)_2$ solution?

$$2KCl_{(aq)} + Pb(NO_3)_{2(aq)} \longrightarrow PbCL_{2(s)} + 2KNO_{3(aq)}$$

Label the balanced equation with the given and required quantities.

$$2KCl_{(aq)} + Pb(NO_3)_{2(aq)} \longrightarrow PbCL_{2(s)} + 2KNO_{3(aq)}$$

0.250 M          0.300 M
  ? L            165 mL = .165 L

The mole map in section 12.9 can be summarized as

$$.165 \text{ L Pb(NO}_3)_2 \text{ solution} \xrightarrow{(M)} ? \text{ mol Pb(NO}_3)_2 \xrightarrow{(mole\ ratio)} ? \text{ mol KCl} \xrightarrow{(M)} ? \text{ L KCl solution}$$

$$.165 \text{ L Pb(NO}_3)_2 \text{ solution} \times \frac{? \text{ mol Pb(NO}_3)_2}{? \text{ L solution}} \times \frac{? \text{ mol KCl}}{? \text{ mol Pb(NO}_3)_2} \times \frac{? \text{ L solution}}{? \text{ mol KCl}} =$$

The first conversion makes use of the molarity of the $Pb(NO_3)_2$ solution,

$$0.300 \text{ M Pb(NO}_3)_2 = \frac{0.300 \text{ mol Pb(NO}_3)_2}{1 \text{ L solution}}$$

$$.165 \text{ L Pb(NO}_3)_2 \text{ solution} \times \frac{0.300 \text{ mol Pb(NO}_3)_2}{1 \text{L solution}} \times \frac{? \text{ mol KCl}}{? \text{ mol Pb(NO}_3)_2} \times \frac{? \text{ L solution}}{? \text{ mol KCl}} =$$

The second conversion utilizes the coefficients of the balanced chemical equation.

$$.165 \text{ L Pb(NO}_3)_2 \text{ solution} \times \frac{0.300 \text{ mol Pb(NO}_3)_2}{1 \text{ L solution}} \times \frac{2 \text{ mol KCl}}{1 \text{ mol Pb(NO}_3)_2} \times \frac{? \text{ L solution}}{? \text{ mol KCl}} =$$

The third conversion makes use of the molarity of the KCl solution,

$$0.250 \text{ M KCl} = \frac{0.250 \text{ mol KCl}}{1 \text{ L solution}}$$

In this instance mol KCl $\longrightarrow$ L solution, so the ratio will be inverted.  The placement of the units to cancel mol KCl requires this arrangement.  Be sure to write 0.250 with the mol unit, and 1 with the L unit.

$$.165 \text{ L Pb(NO}_3)_2 \text{ solution} \times \frac{0.300 \text{ mol Pb(NO}_3)_2}{1 \text{ L solution}} \times \frac{2 \text{ mol KCl}}{1 \text{ mol Pb(NO}_3)_2} \times \frac{1 \text{ L solution}}{0.250 \text{ mol KCL}} =$$

$$= 0.396 \text{ L KCl solution}$$
$$= 396 \text{ mL KCl solution}$$

40.  Calculate the molarity of $CuSO_4$ if 0.195g of CuS are formed by complete reaction of 750.0 mL of $CuSO_4$ solution.

Label the balanced chemical equation.

$$Na_2S_{(aq)} + CuSO_{4(s)} \longrightarrow Na_2SO_{4(aq)} + CuS_{(s)}$$

$$\quad\quad\quad\quad\quad 750.0 \text{ mL} \quad\quad\quad\quad\quad\quad\quad\quad\quad 0.195g$$
$$\quad\quad\quad\quad\quad\quad ? \text{ M}$$

$$\text{M CuSO}_4 = \frac{\text{moles CuSO}_4}{\text{L of CuSO}_4 \text{ solution}}$$

The volume of $CuSO_4$ solution is given.  To find M $CuSO_4$, first find moles    from a stoichiometric calculation.

$0.195g$ CuS $\xrightarrow{CuSO_4}$ ? mol CuS $\longrightarrow$ ? mol $CuSO_4$

$0.195g$ CuS x $\dfrac{1 \text{ mol CuS}}{95.6g \text{ CuS}}$ x $\dfrac{1 \text{ mol } CuSO_4}{1 \text{ mol CuS}} = 2.04 \times 10^{-3}$ mol $CuSO_4$

M $Na_2S = \dfrac{2.04 \times 10^{-3} \text{ mol } CuSO_4}{.7500 \text{ L solution}}$

$= \dfrac{2.72 \times 10^{-3} \text{ mol } CuSO_4}{1 \text{ L solution}} = 2.72 \times 10^{-3}$ M $CuSO_4$

***Practice Problem E

Sodium hypochlorite, NaClO, is used as a bleaching agent. Sodium hypochlorite can be prepared by the reaction

$Cl_2{}_{(g)}$ + 2NaOH$_{(aq)} \longrightarrow$ NaClO$_{(aq)}$ + NaCl$_{(aq)}$ + $H_2O_{(l)}$

How many grams of $Cl_2$ are required to react with 1.00 L of 5.00 M NaOH?

45.  Calculate the molality of each of the following solutions.

a.  2.30g of $C_2H_5OH$ dissolved in 400.0g of water

To calculate molality,

$$m = \dfrac{\text{mol solute}}{\text{kg of solvent}}$$

first calculate mol $C_2H_5OH$.

$2.30g$ $C_2H_5OH$ x $\dfrac{1 \text{ mol } C_2H_5OH}{46.06g \text{ } C_2H_5OH} = 4.99 \times 10^{-2}$ mol $C_2H_5OH$

The mass of the solvent, water, must be expressed in kg.

$400.0g$ $H_2O$ x $\dfrac{1 \text{ kg}}{10^3 g} = .400$ kg $H_2O$

The molality can now be calculated.

$m = \dfrac{4.99 \times 10^{-2} \text{ mol } C_2H_5OH}{.400 \text{ kg } H_2O}$

$= \dfrac{1.25 \times 10^{-1} \text{ mol } C_2H_5OH}{1 \text{ kg } H_2O}$

$= 1.25 \times 10^{-1}$ m $C_2H_5OH$

Notice that in molality, the concentration expression compares quantity of solute to solvent.  The mass of water is not the mass of the whole solution.

Consult Appendix C in the text for answers to parts b, c, and d.

59.  A solution of 14.5g of fructose (a sugar found in fruits and honey) dissolved in 150.0g of water has a freezing point of -1.0°C.  What is the molar mass of fructose.

Molar mass can be calculated using the rearranged freezing point lowering equation.

$$\text{Molar mass fructose} = K_f\left(\frac{\text{g solute}}{\Delta t_f}\right)\left(\frac{1}{\text{kg solvent}}\right)$$

The freezing point depression constant for the solvent water is given in Table 12-6.  $K_f$ for water = $\dfrac{1.86°C \cdot \text{kg } H_2O}{\text{mol solute}}$

Pure water has a freezing point of 0°C (Table 12-6).  The fructose solution has a freezing point of -1.0°C.

$\Delta t_f$ = normal freezing point-freezing point of solution

   = 0°C-(-1.0°C)
   = 1.0°C

Convert the g of $H_2O$ to kg of $H_2O$ so that the units will be correct.

$$150.0\text{g } H_2O \times \frac{1 \text{ kg } H_2O}{10^3\text{g } H_2O} = .1500 \text{ kg } H_2O$$

Substitute the data into the equation above to find molar mass of fructose.

$$\text{molar mass} = \left(\frac{1.86°C \cdot \text{kg } H_2O}{\text{mol fructose}}\right)\left(\frac{14.5\text{g fructose}}{1.0°C}\right)\left(\frac{1}{.1500 \text{ kg } H_2O}\right)$$

$$= \frac{180 \text{ g fructose}}{\text{mol}}$$

***Practice Problem F

When 0.500g of vitamin K is dissolved in 10.0g of camphor, the freezing point of the solution is 175.1°C.  Calculate the molar mass of vitamin K.  (Consult Table 12-6 for $K_f$ and freezing point values.)

## SELF-TEST

Circle the correct answer in the following multiple choice questions.

1. Which of the following statements is true for water?

   a. Molecules in the solid state have fewer hydrogen bonds than in the liquid state.
   b. Water's relatively high melting point and boiling point can be attributed to its strong London dispersion forces.
   c. The solid state of water is more dense than the liquid state.
   d. Water is a bent, polar molecule.

2. A solution

   a. can be clear or cloudy.
   b. must be colorless.
   c. has two components, solute and solvent.
   d. is a heterogeneous mixture of dissolved ions.

3. Which of the following statements correctly describes solution formation?

   a. Solute particles must be attracted more strongly to solvent molecules than to other solute particles.
   b. When water molecules surround solute particles, the process is called aquation.
   c. HCl is attracted to water by hydrogen bonding.
   d. Ionic compounds can dissolve only in ionic solvents.

4. The solubility of NaCl is 36.0g/100 mL $H_2O$.  The solubility of AgCl is $1.50 \times 10^{-4}$g/100 mL $H_2O$.  Which of the following is a true statement?

   a. AgCl is a soluble compound.
   b. when 40.0g of NaCl is added to 100 mL of $H_2O$, a saturated solution results.
   c. 1.5g of AgCl will dissolve in 1.00 L of water.
   d. The solubility of AgCl should decrease with an increase in temperature.

5. A solution is prepared by dissolving 55.0g of NaClO in enough water to make 125 mL of solution.  The molarity of the sodium hypochlorite solution is

   a. 4.40 M NaClO
   b. 5.91 M NaClO
   c. 0.924 M NaClO
   d. 10.8 M NaClO

6. How many grams of NaCl are dissolved in 500.0 mL of a solution which is 1.25 M NaCl?

   a. 36.5 g NaCl
   b. 23.4 g NaCl
   c. 93.5 g NaCl
   d. 73.1 g NaCl

7.      Calculate the %(w/w) concentration of a solution prepared by
        adding 25.0g NaOH to water.  The resulting solution weighed
        118.5g

        a.   .523%(w/w)  NaOH
        b.   4.74%(w/w)  NaOH
        c.   26.7%(w/w)  NaOH
        d.   21.1%(w/w)  NaOH

8.      What volume of a 3.5%(w/v) $CH_3OH$ solution can be prepared
        from 10.0g $CH_3OH$?

        a.   35.0 mL
        b.   286 mL
        c.   .350 L
        d.   2.87 mL

9.      What volume of a 3.50 M KOH solution is needed to react
        completely with 200.0 mL of a 6.50 M $H_2SO_4$ solution?

$$H_2SO_{4(aq)} + 2KOH_{(aq)} \longrightarrow K_2SO_{4(aq)} + 2H_2O_{(l)}$$

        a.   372 mL
        b.   743 mL
        c.   9.10 L
        d.   186 mL

10.     Calculate the concentration of a solution prepared by
        diluting 100.0 mL of a 15.0%(w/v) NaCl solution to 325.0 mL.

        a.   48.8%
        b.   .0205%
        c.   4.62%
        d.   1.50%

---

## ANSWERS TO PRACTICE PROBLEMS

Practice Problem A

a.   mol NaOH = 25.0g NaOH x $\dfrac{1 \text{ mol NaOH}}{40.0 \text{g NaOH}}$

     = .625 mol NaOH

     M NaOH = $\dfrac{.625 \text{ mol NaOH}}{5.00 \text{ L NaOH solution}}$ = .125 M NaOH

b.   Given information:   200.0 mL solution
     Conversion factor:   $\dfrac{1.25 \text{ mol NaOH}}{\text{L solution}}$

     200.0 mL solution $\longrightarrow$ L solution $\longrightarrow$ mol NaOH $\longrightarrow$ g NaOH

200.0 mL solution x $\dfrac{10^{-3} \text{L}}{1 \text{ L}}$ x $\dfrac{1.25 \text{ mol NaOH}}{1 \text{ L solution}}$ x $\dfrac{40.0 \text{g NaOH}}{1 \text{ mol NaOH}}$ = 10.0g NaOH

c.   Given information:   50.0g NaOH
     Conversion factor:   $\dfrac{1.25 \text{ mol NaOH}}{\text{L solution}}$

50.0g NaOH $\longrightarrow$ mol NaOH $\longrightarrow$ 1 solution

$50.0\text{g NaOH} \times \dfrac{1 \text{ mol NaOH}}{40.0\text{g NaOH}} \times \dfrac{1 \text{ L solution}}{1.25 \text{ mol NaOH}} = 1.00 \text{ L solution}$

## Practice Problem B

a.   $\%(w/v) \text{ NaCl} = \dfrac{2.30\text{g NaCl}}{250.0 \text{ mL}} \times 100$

$= 0.920\% (w/v) \text{ NaCl}$

b.   Given information:   425 mL NaCl solution
     Conversion factor:   $\dfrac{0.920\text{g NaCl}}{100 \text{ mL solution}}$

425 mL solution $\xrightarrow{\%(w/v)}$ g NaCl

$425 \text{ mL solution} \times \dfrac{0.920\text{g NaCl}}{100 \text{ mL solution}} = 3.91\text{g NaCl}$

c.   Given information:   5.00g NaCl
     Conversion factor:   $\dfrac{0.920\text{g NaCl}}{100 \text{ mL solution}}$

5.00g NaCl $\xrightarrow{\%(w/v)}$ mL solution

$5.00\text{g NaCl} \times \dfrac{100 \text{ mL solution}}{0.920\text{g NaCl}} = 543 \text{ mL solution}$

## Practice Problem C

a.   $\%(v/v) \text{ CH}_3\text{CH}_2\text{OH} = \dfrac{50.0 \text{ mL CH}_3\text{CH}_2\text{OH}}{125.0 \text{ mL solution}} \times 100$

$= 40.0\% (v/v) \text{ CH}_3\text{CH}_2\text{OH}$

b.   $\%(w/v) \text{ CH}_3\text{CH}_2\text{OH} = \dfrac{\text{g CH}_3\text{CH}_2\text{OH}}{\text{mL solution}} \times 100$

Use the density of ethyl alcohol to find the mass of 50.0 mL of ethyl alcohol.

$50.0 \text{ mL} \times \dfrac{0.789\text{g}}{1 \text{ mL}} = 39.5\text{g}$

$\%(w/v) \text{ CH}_3\text{CH}_2\text{OH} = \dfrac{39.5\text{g CH}_3\text{CH}_2\text{OH}}{125.0 \text{ mL solution}} \times 100$

$= 31.6\% (w/v) \text{ CH}_3\text{CH}_2\text{OH}$

c.   $\%(w/w) \text{ CH}_3\text{CH}_2\text{OH} = \dfrac{\text{g CH}_3\text{CH}_2\text{OH}}{\text{g solution}} \times 100$

Use the density of the solution to find the mass of 125.0 mL of the solution.

$125.0 \text{ mL} \times \dfrac{0.950\text{g}}{1 \text{ mL}} = 119\text{g}$

Use the mass of $CH_3CH_2OH = \dfrac{39.5g\ CH_3CH_2OH}{119g\ solution} \times 100$

$= 33.2\%\ (w/w)\ CH_3CH_2OH$

Notice that the different percent concentration expressions differ in value for the $CH_3CH_2OH$ solution.

d.  The volume of solution is given, and a mass of solute is required.  The appropriate conversion factor to use is $\%(w/v)$.

Given information:  200.0 mL solution
Conversion factor:  $\dfrac{31.6g\ CH_3CH_2OH}{100\ mL\ solution}$

200.0 mL solution $\xrightarrow{\%(w/v)}$ g $CH_3CH_2OH$

200.0 mL solution $\times\ \dfrac{31.6g\ CH_3CH_2OH}{100\ mL\ solution} = 63.2g\ CH_3CH_2OH$

## Practice Problem D

a.  State the problem.

$C_c = 7.50\%\,(w/v)$　　　　　$C_d = ?$
$V_c = 100.0\ ml$　　　　　$V_d = 250.0\ mL$

$$C_d = C_c\left(\frac{V_c}{V_d}\right)$$

$$= 7.50\%\,(w/v)\left(\frac{100.0\ ml}{250.0\ ml}\right)$$

$$= 3.00\%\,(w/v)$$

b.  State the problem.

$C_c = 15.0\%\,(v/v)$　　　　　$C_d = 10.0\%\,(v/v)$
$V_c = 75.0\ mL$　　　　　$V_d = ?$

$$V_d = V_c\left(\frac{C_c}{C_d}\right)$$

$$= 75.0\ mL\left(\frac{15.0\%\,(v/v)}{10.0\%\,(v/v)}\right)$$

$$= 113\ mL$$

## Practice Problem E

Label the balanced chemical equation.

$$Cl_{2\,(g)} + 2NaOH_{(aq)} \longrightarrow NaClO_{(aq)} + NaCl_{(aq)} + H_2O_{(l)}$$

?g　　　1.00 L
　　　　5.00 M

1.00 L NaOH solution $\longrightarrow$ ? mol NaOH $\longrightarrow$ ? mol $Cl_2$ $\longrightarrow$ ? g $Cl_2$

1.00 L NaOH solution $\times\ \dfrac{5.00\ mol\ NaOH}{1\ L\ solution} \times\ \dfrac{1\ mol\ Cl_2}{2\ mol\ NaOH} \times\ \dfrac{70.9g\ Cl_2}{1\ mol\ Cl_2} =$

$177g\ Cl_2$

Practice Problem F

$$\text{Molar mass of Vitamin K} = K_f\left(\frac{g\ \text{solute}}{\Delta t_f}\right)\left(\frac{1}{kg\ \text{solvent}}\right)$$

From Table 12-6:

$$K_f\ \text{camphor} = \frac{39.7°C \cdot kg\ \text{camphor}}{\text{mol solute}}$$

Freezing point camphor = 179.5°C

Calculate $\Delta t_f$

$$\Delta t_f = 179.5°C - 175.1°C = 4.4°C$$

The solvent must be expressed in kg.

$$10.0g\ \text{camphor} \times \frac{1\ kg}{10^3 g} = .0100\ kg\ \text{camphor}$$

Substitute the data into the equation above to find molar mass of Vitamin K.

$$\text{molar mass Vitamin K} = \left(\frac{39.7°C \cdot kg\ \text{camphor}}{\text{mol Vitamin K}}\right)\left(\frac{0.500g\ \text{Vitamin K}}{4.4°C}\right)\left(\frac{1}{.0100\ kg\ \text{camphor}}\right)$$

$$= 451\ \frac{g}{\text{mol}}\ \text{Vitamin K}$$

---

ANSWERS TO SELF-TEST

1.  d
2.  c
3.  a
4.  b
5.  b
6.  a
7.  d
8.  b
9.  b
10. c

# CHEMICAL EQUILIBRIUM AND REACTION RATES

# CHAPTER 13



**13.1 Reversible Reactions**

Reversible reactions are reactions that proceed in both the forward and reverse directions simultaneously.

A reaction written with two opposing arrows (A $\rightleftharpoons$ B) is reversible.

Reversible reactions progress toward a state of dynamic equilibrium in which the rate of the forward reaction is equal to the rate of the reverse reaction (Figure 13-1).

**13.2 Rates of Chemical Reactions**

The rate of a chemical reaction is the number of formula units of a product formed in a given period of time.

Chemical kinetics is the study of reaction rates.

**13.3 Activation Energy**

The reaction pathway is a hypothetical route by which the reactants are converted to products in a chemical reaction.

An energy barrier exists in every reaction pathway. The amount of energy needed to cross the barrier is called the activation energy, $E_a$ (Figure 13-3).

Reactions proceed more rapidly when $E_a$ is low than when $E_a$ is high.

When the reactant formula units acquire energy equal to the activation energy, their structures change.  These structures are called activated complexes or transition states.

## 13.4   Factors Affecting Reaction Rates

Higher temperature increases the vibration of atoms within molecules and the frequency of collisions between molecules.  Thus, reaction rate is faster at higher temperatures.

A higher concentration of one or more reactants results in more frequent collisions and thus faster reaction rates.

A decrease in the volume of gaseous reactants increases the concentration and thus the rate of the reaction.

Reactions rates are expressed in units of concentration change per unit of time, such as

$$\frac{\frac{moles}{liter}}{min} = mol\ L^{-1}\ min^{-1}$$

or

$$\frac{\frac{moles}{liter}}{sec} = mol\ L^{-1}\ sec^{-1}$$

The rate of a unimolecular reaction (A $\longrightarrow$ B) depends on the concentration of only one reactant.

The rate equation or rate law for a unimolecular reaction is described by the following expression

$$rate = k[A]$$

where:   [A] is the concentration of A (the brackets denote concentration in mol/L), and
k is the rate constant or constant of proportionality.

The rate of a bimolecular reaction (A + B $\longrightarrow$ C) depends on the concentration of two reactants.

The rate law for a bimolecular reaction is

$$rate = k[A][B]$$

The rate constant k is characteristic of a particular reaction and only a temperature change will change its value.

The activation energy, $E_a$, and the rate constant, k, are unique to each reaction because their values depend on the nature of the reactants.  Reactions between molecules are usually slower than reactions between ions.

A catalyst affects the rate of a chemical reaction without undergoing any permanent change.

It is probable that a catalyst speeds up a reaction by providing a new reaction pathway in which $E_a$ is lower (Figure 13-4).

## 13.5  Equilibrium Constants

At equilibrium, the ratio of molar concentrations of product(s) and reactant(s) is constant at a specific temperature.  The constant is called the equilibrium constant, $K_{eq}$.

For the general reaction,

$$aA + bB + cC + \ldots \rightleftharpoons dD + eE + fF + \ldots$$

$$K_{eq} = \frac{[D]^d[E]^e[F]^f + \ldots}{[A]^a[B]^b[C]^c + \ldots}$$

The units of $K_{eq}$ will vary, depending on the exponents in the equilibrium constant expression.

## 13.6  Applications of Equilibrium Constants

$K_{eq}$ provides information about the relative amounts of product(s) and reactant(s) that exist at equilibrium.

When $K_{eq} < 1$, the equilibrium favors reactants, and the position of equilibrium lies on the left.

When $K_{eq} > 1$, product formation is favored, and the position of equilibrium lies on the right.

## 13.7  LeChatelier's Principle

LeChatelier's principle states that if a chemical equilibrium is disturbed, the system will readjust to offset the disturbance and return to equilibrium.

Changes in reactant or product concentrations will change the forward and reverse reaction rates, but not change the value of $K_{eq}$ (as long as temperature is held constant).  The equilibrium will shift in the direction indicated below in response to changes in concentration.

$$\text{reactants} \rightleftharpoons \text{products}$$

$$\text{increased [reactants]} \longrightarrow$$

$$\longleftarrow \text{increased [products]}$$

In chemical equilibria involving one or more gases, a change in volume is equivalent to a change in concentration of the gas(es).  A decrease in volume shifts the equilibrium to the side of the balanced equation with the fewer moles of gas.  If the number of moles of gaseous reactant and product are equal, a change in volume (or applied pressure) will have no effect on equilibrium.

Changes in the temperature of a system will change the forward and reverse reaction rates and the value of $K_{eq}$.  Temperature changes will shift the equilibrium in the direction indicated below.

$$\text{reactants} \rightleftharpoons \text{products} + \text{heat}$$
$$\xleftarrow{\hspace{1cm}} \text{increased temperature}$$

$$\text{reactants} + \text{heat} \rightleftharpoons \text{products}$$
$$\text{increased temperature} \longrightarrow$$

## 13.8    Energy and Chemical Change

The heat absorbed or released by a chemical reaction at constant pressure is called the enthalpy change, $\Delta H$.

$$\Delta H = \text{heat content}_{products} - \text{heat content}_{reactants}$$

The enthalpy changes in exothermic and endothermic reactions are summarized below.

|  | endothermic reaction | exothermic reaction |
|---|---|---|
| general reaction | reactants+heat $\rightleftharpoons$ products | reactants $\rightleftharpoons$ products+heat |
| direction of heat flow | heat absorbed | heat released |
| $heat_r$ vs. $heat_p$ | $heat_{reactants} < heat_{products}$ | $heat_{reactants} > heat_{products}$ |
| sign of $\Delta H$ | $+ \Delta H$ | $- \Delta H$ |

A calorimeter (Figure 13-6) measures enthalpy changes.  Heat absorbed or released by a reaction is reflected by a change in water temperature.

## 13.9    The Relationship Between Chemical Equilibrium and Reaction Rates

When a reaction is at equilibrium, the rates of the two opposing reactions are equal:

$$\text{rate of forward reaction} = \text{rate of reverse reaction}$$

For the reaction

$$A \rightleftharpoons B$$
$$k_f[A] = k_r[B]$$

where $k_f$ is the rate constant for the forward reaction and $k_r$ is the rate constant for the reverse reaction.  Rearrangement of terms gives

$$\frac{k_f}{k_r} = \frac{[B]}{[A]}$$

Since

$$K_{eq} = \frac{[B]}{[A]}, \text{ then}$$

$$K_{eq} = \frac{k_f}{k_r}$$

All equilibrium constants are ratios of rate constants.

---

SOLUTIONS TO SELECTED TEXT STUDY QUESTIONS AND PROBLEMS
and
PRACTICE PROBLEMS

---

11. For the general reaction

$$A+B \longrightarrow C+D$$

what is the effect on the rate of

a. doubling the concentration of A?

The rate law for this equation is

$$rate = k[A][B]$$

Assuming that the concentration of A and B was 1 mol/L, the original rate was

$$rate = k\left(\frac{1\ mol}{L}\right)\left(\frac{1\ mol}{L}\right)$$

$$= 1\ k\ \frac{mol^2}{L^2} = 1\ k\ mol^2/L^2$$

If the concentration of A is doubled, then the new concentration is 2 mol/L, and the new rate is

$$rate = k(2\ mol/L)(1\ mol/L)$$
$$= 2\ k\ mol^2/L^2$$

Thus, doubling the concentration of A doubles the rate. Notice that different values selected for [A] and [B] will not affect the outcome. Consider the problem as a ratio of rates, the initial rate ($rate_1$) and the rate of the reaction after [A] is doubled ($rate_2$).

$$\frac{rate_2}{rate_1} = \frac{k[2A][B]}{k[A][B]}$$

Since k is constant at constant temperature, and the concentration of B is not changed, the ratio of rates can be simplified.

$$\frac{rate_2}{rate_1} = \frac{\cancel{k}[2A]\cancel{[B]}}{\cancel{k}[A]\cancel{[B]}}$$

$$\frac{rate_2}{rate_1} = 2,\ or$$

$$rate_2 = 2(rate_1)$$

This relationship is true for any values selected initially for [A] and [B].

b. doubling the concentration of B?

Assume:

$$[A] = 1\ mol/L$$
$$[B] = 1\ mol/L$$

$$rate = k(1\ mol/L)(1\ mol/L)$$
$$= 1\ k\ mol^2/L^2$$

If the concentration of B is doubled,

$$\text{rate} = k(1 \text{ mol/L})(2 \text{ mol/L})$$
$$= 2 \text{ k mol}^2/\text{L}^2$$

Thus, doubling the concentration of B doubles the rate.

c.   doubling the concentration of A and B?

Assume:

$$[A] = 1 \text{ mol/L}$$
$$[B] = 1 \text{ mol/L}$$

$$\text{rate} = k(1 \text{ mol/L})(1 \text{ mol/L})$$
$$= 1 \text{ k mol}^2/\text{L}^2$$

If the concentration of A and B are doubled,

$$\text{rate} = k(2 \text{ mol/L})(2 \text{ mol/L})$$
$$= 4 \text{ k mol}^2/\text{L}^2$$

Thus, doubling the concentration of both A and B quadruples the rate.

***Practice Problem A

For the reaction,

$$2 \text{ NO}_{(g)} + \text{O}_{2(g)} \longrightarrow 2 \text{ NO}_{2(g)}$$

a.   Write the rate equation.

b.   If the rate of formation of $NO_2$ is 7.10 mol $L^{-1}$ sec$^{-1}$, under the following conditions,

$$[NO] = 0.0010 \text{ mol} \cdot L^{-1}$$
$$[O_2] = 0.0010 \text{ mol} \cdot L^{-1}$$

what is the value of the rate constant, k?

17.   Calculate the equilibrium constant for the reaction

$$\text{H}_{2(g)} + \text{I}_{2(g)} \rightleftharpoons 2 \text{ HI}_{(g)}$$

on the basis of the following equilibrium concentrations:

$$[H_2] = 0.90 \text{ mol } L^{-1}$$
$$[I_2] = 0.40 \text{ mol } L^{-1}$$
$$[HI] = 0.60 \text{ mol } L^{-1}$$

$$K_{eq} = \frac{[HI]^2}{[H_2][I_2]}$$

$$= \frac{(0.60 \text{ mol} \cdot L^{-1})^2}{(0.90 \text{ mol} \cdot L^{-1})(0.40 \text{ mol} \cdot L^{-1})}$$

$$= \frac{(0.60 \text{ mol} \cdot L^{-1})(0.60 \text{ mol} \cdot L^{-1})}{(0.90 \text{ mol} \cdot L^{-1})(0.40 \text{ mol} \cdot L^{-1})}$$

$$= 1.0$$

$K_{eq}$ has no units in this problem.

***Practice Problem B

Nitrogen and hydrogen gases are reacted at 500°C, according to the following reaction.

$$N_{2(g)} + 3 H_{2(g)} \rightleftharpoons 2 NH_{3(g)}$$

When equilibrium is established,

$$[N_2] = 0.600 \text{ mol} \cdot L^{-1}$$
$$[H_2] = 0.420 \text{ mol} \cdot L^{-1}$$
$$[NH_3] = 0.0596 \text{ mol} \cdot L^{-1}$$

a.  Calculate $K_{eq}$ for the reaction at 500°C.

b.  Write the balanced chemical equation using arrows of different lengths to indicate the position of equilibrium.

c.  If the concentration of $N_2$ increases, in what direction does the equilibrium shift?

25.    Predict the concentration changes that would occur and the direction in which each of the following equilibria would shift in response to the indicated changes:

a.   $N_2(g) + O_2(g) \rightleftharpoons 2NO(g)$

To overcome the stress of additional $O_2$ the reaction shifts to the right to consume $O_2$.  In the process, $[N_2]$ will decrease and $[NO]$ will increase.  This is confirmed by examining the equilibrium expression.

$$K_{eq} = \frac{[NO]^2}{[N_2][O_2]}$$

In order for $K_{eq}$ to remain constant when $[O_2]$ increases, $[NO]$ must increase and $[N_2]$ must decrease.  When compared to the chemical reaction, this can only be accomplished if the equilibrium shifts to the right.

Consult Appendix C in the text for answers to parts b, c, and d.

28.    A mixture of $H_2(g)$, $O_2(g)$, and $H_2O(g)$ is in equilibrium in a closed container.  Assuming that the formation of water,

$$2H_2(g) + O_2(g) \rightleftharpoons 2H_2O(g)$$

is exothermic, predict the effect of each of the following changes on the concentration of $H_2(g)$:

c.   increasing the temperature of the system.

Write the equation showing heat as a reactant or product. Exothermic reactions release heat (see section 13.8) so heat is a product

$$2H_2(g) + O_2(g) \rightleftharpoons 2H_2O(g) + heat$$

To overcome the stress of a higher temperature, the position of equilibrium shifts to consume heat.  Shift to the right produces more heat, a shift to the left consumes heat.  Thus the temperature increase causes a shift to the left in the reaction above.  Thus $[H_2]$ and $[O_2]$ increase  while $[H_2O]$ decreases.

Consult Appendix C in the text for answers to a and b.

***Practice Problem C

The reaction

$$CH_4(g) + 2H_2S(g) \rightleftharpoons CS_2(g) + 4H_2(g)$$

is endothermic.  How will the position of equilibrium be affected by the following changes?

a.   an increase in temperature

b.   a decrease in the volume of the container

c.   removal of $H_2{}_{(g)}$

---

SELF-TEST

Circle the correct answer in the following multiple choice questions.

1.   Which of the following statements is true for activation energy, $E_a$?

   a.   When $E_a$ is high, the rate of reaction is high.
   b.   When $E_a$ is high, formation of product is favored.
   c.   When $E_a$ is low, the rate of reaction is high.
   d.   $E_a$ is a measure of the ratio of reactants to products.

2.   The rate of a reaction will increase

   a.   when the reactant concentration increases.
   b.   when the molecules slow down.
   c.   when molecules collide with less force.
   d.   when temperature decreases.

3.   Which reaction is described by the following rate law?

$$\text{rate} = k[H_2O_2]^2$$

   a.   $(H_2O_2)_2 \longrightarrow 2H_2O + O_2$

   b.   $2H_2O + O_2 \longrightarrow 2H_2O_2$

   c.   $H_2 + O_2 \longrightarrow H_2O_2$

   d.   $2H_2O_2 \longrightarrow 2H_2O + O_2$

4.   The equilibrium constant expression,

$$K_{eq} = \frac{[N_2][CO_2]^2}{[NO]^2[CO]^2}$$

is written for the following reaction

   a.   $N_2 + 2CO_2 \longrightarrow 2NO + 2CO$
   b.   $2N + 2CO_2 \longrightarrow NO_2 + CO_2$
   c.   $2NO + 2CO \rightleftharpoons N_2 + 2CO_2$
   d.   $NO_2 + CO_2 \longrightarrow N_2 + 2CO_2$

5.  The position of equilibrium for the following reaction lies to the left (formation of products is not favored).

$$2\ CH_{4(g)} \rightleftharpoons C_2H_{6(g)} + H_{2(g)}$$

The value for $K_{eq}$ is

a.  $9.5 \times 10^{-13}$

b.  $3.6 \times 10^{20}$

c.  $1.00$

d.  $0$

6.  The equilibrium will shift to the right for the following reaction

$$CO_{(g)} + 2H_{2(g)} \rightleftharpoons CH_3OH + heat$$

when

a.  $[H_2]$ is decreased.
b.  the temperature is raised.
c.  $[CO]$ is increased.
d.  $[CH_3OH]$ is increased.

7.  A change in pressure on the following reaction at equilibrium

$$CH_{4(g)} + H_2O_{(g)} \rightleftharpoons CO_{(g)} + 3H_{2(g)}$$

a.  will cause no change in equilibrium.
b.  will shift the equilibrium to the right if pressure increases.
c.  will shift the equilibrium to the left if pressure decreases.
d.  will shift the equilibrium to the right if pressure decreases.

8.  If the concentration of A is doubled in the following reaction

$$2A + B \longrightarrow 2C + 2D$$

how does the rate of the reaction change?

a.  doubles
b.  triples
c.  quadruples
d.  no change

9.  The rate of formation of C is $2.0 \times 10^{-4}$ mol $L^{-1}$ $sec^{-1}$ in the following reaction

$$A+B \longrightarrow C$$

when [A] is 0.10 mol$\cdot$$L^{-1}$ and [B] is 0.10 mol$\cdot$$L^{-1}$.  The value of the rate constant, k, is

a.  $2.0 \times 10^{-3}$ $sec^{-1}$
b.  $2.0 \times 10^{-2}$ L$\cdot$mol$^{-1}$ $sec^{-1}$
c.  $5.0 \times 10^{1}$ mol sec $L^{-1}$
d.  $2.0 \times 10^{-6}$ mol $L^{-1}$ $sec^{-1}$

10. Sulfur dioxide and oxygen are reacted at 1500K, according to the following reaction

$$2SO_{2\,(g)} + O_{2\,(g)} \rightleftharpoons 2SO_{3\,(g)}$$

when equilibrium is established

$$[SO_2] = 0.344 \text{ M}$$
$$[O_2] = 0.172 \text{ M}$$
$$[SO_3] = 0.056 \text{ M}$$

The $K_{eq}$ for the reaction at 1500K is

a. $.946 \text{ L mol}^{-1}$
b. $.154 \text{ L mol}^{-1}$
c. $6.49 \text{ mol} \cdot \text{L}^{-1}$
d. $1.06 \text{ mol} \cdot \text{L}^{-1}$

---

## ANSWERS TO PRACTICE PROBLEMS

### Practice Problem A

a. $\text{rate} = k[NO]^2[O_2]$

b. $k = \dfrac{\text{rate}}{[NO]^2[O_2]} = \dfrac{7.10 \text{ mol} \cdot \text{L}^{-1} \text{ sec}^{-1}}{(0.0010 \text{ mol L}^{-1})^2(0.0010 \text{ mol L}^{-1})}$

$$= 7.10 \times 10^9 \text{ mol}^{-2} \text{ L}^2 \text{ sec}^{-1}$$

### Practice Problem B

a. $K_{eq} = \dfrac{[NH_3]^2}{[N_2][H_2]^3} = \dfrac{(0.0596)^2}{(0.600)(0.420)^3}$

$$= 0.0799$$

b. $N_{2\,(g)} + 3H_{2\,(g)} \rightleftharpoons 2NH_{3\,(g)}$

The equilibrium favors reactants, since $K_{eq} \ll 1.0$.

c. To the right.

### Practice Problem C

$$CH_{4\,(g)} + 2H_2S_{(g)} + \text{heat} \rightleftharpoons CS_{2\,(g)} + 4H_{2\,(g)}$$

a. Equilibrium shifts to the right.
b. Equilibrium shifts to the left.
c. Equilibrium shifts to the right

---

## ANSWERS TO SELF-TEST

| | | | |
|---|---|---|---|
| 1. | c | 6. | c |
| 2. | a | 7. | d |
| 3. | d | 8. | c |
| 4. | c | 9. | b |
| 5. | a | 10. | b |

# ACIDS, BASES, AND SALTS

CHAPTER

**14**

## 14.1 Concepts of Acids and Bases

The Arrhenius theory defines acids as substances that produce $H^+$ in water and bases as substances that produce $OH^-$ in water.

The following general equations illustrate the Arrhenius theory.

$$HA \xrightarrow{H_2O} H^+_{(aq)} + A^-_{(aq)}$$

where HA represents any acid and $A^-$ represents the corresponding anion.

$$MOH \xrightarrow{H_2O} M^+_{(aq)} + OH^-_{(aq)}$$

where M represents any metallic element and $M^+$ represents the corresponding cation.

A more general theory of acids and bases, the Bronsted-Lowry theory, defines an acid as a substance that donates a proton ($H^+$) and a base as a substance that accepts a proton ($H^+$)

The following general equation illustrates the Bronsted-Lowry theory.

$$H\text{-}ACID + BASE \rightleftharpoons H\text{-}BASE^+ + ACID^-$$
$$(acid_1) \quad (base_1) \quad (acid_2) \quad (base_2)$$

When the reactant acid transfers its proton to the base, the products formed are also an acid and a base. H-BASE$^+$ is identified as an acid because it donates a proton in the reaction to the left. ACID$^-$ is identified as a base because it accepts a proton in the reaction to the left.

When H-ACID loses a proton, the base ACID$^-$ forms. ACID$^-$ is the conjugate base of H-ACID.

When BASE gains a proton, the acid H-BASE$^+$ forms. H-BASE$^+$ is the conjugate acid of BASE. Each pair of bracketed substances is an acid-base conjugate pair.

Most common acids are either molecules or cations and most bases are molecules or anions.

Substances (like $H_2O$) that can donate or accept protons are said to be amphoteric. When water accepts a proton, the hydronium ion forms (Figure 14-2).

$$H_2O + H^+ \longrightarrow H_3O^+$$

When water donates a proton, the hydroxide ion forms.

$$H_2O \longrightarrow H^+ + OH^-$$

## 14.2   Properties of Acids and Bases

An acid-base indicator is a substance that has one color in acid solution and another color in basic solution.

In aqueous solution, acids have a sour taste, turn blue litmus red, neutralize bases and react with certain metals.

Aqueous bases have a bitter taste and slippery feeling, turn red litmus blue, and neutralize acids.

## 14.3   Common Acids and Bases

Carbonated beverages contain carbonic acid, $H_2CO_3$, which forms when $CO_2$ dissolves in water:

$$CO_{2(g)} + H_2O_{(1)} \rightleftharpoons H_2CO_{3(aq)}$$

However, only 1% of the $CO_2$ reacts to form $H_2CO_3$. The equilibrium lies far to the left.

Sodium hydroxide, NaOH, and ammonia, $NH_3$, are the most common bases.

## 14.4   Acid Strength

The ionization reaction of a general acid, HA, in water is

$$HA_{(aq)} + H_2O_{(1)} \rightleftharpoons H_3O^+_{(aq)} + A^-_{(aq)}$$

(HA can be a molecule or ion.)

The $K_{eq}$ for the ionization of an acid is $K_{eq} = \dfrac{[H_3O^+][A^-]}{[HA][H_2O]}$

(In most aqueous solutions $[H_2O]$ is nearly 55.5 M, the concentration of pure water.  Thus, $[H_2O]$ is a constant.)

Combining constants, $K_{eq}[H_2O] = \dfrac{[H_3O^+][A^-]}{[HA]}$

a new constant is derived called $K_a$, the acid ionization constant.

When $K_a \geq 1$, the position of equilibrium lies to the right, the acid is almost totally ionized, and the acid is classified as strong (Table 14-3).

When $K_a < 1$, the acid is only slightly ionized, and is classified as weak (Table 14-4).

The weaker an acid, the stronger its conjugate base, and the stronger an acid, the weaker its conjugate base (Table 14-5).

## 14.5   Polyprotic Acids

Polyprotic acids are capable of donating more than one proton from each formula unit.  Polyprotic acids ionize in steps, donating one proton in each step.  Sulfuric acid ionizes in two steps.  Each step in the ionization of a polyprotic acid has its own $K_a$.

The acid strength of a polyprotic acid is related to the electronegativity of the central nonmetallic atom.  The higher the electronegativity of the nonmetal, the stronger the acid.

## 14.6   Base Strength

Soluble metal hydroxides are strong bases.

Bases that do not contain hydroxides are usually weak.  The general reaction for a weak base is

$$B + H_2O_{(1)} \rightleftharpoons BH^+_{(aq)} + OH^-_{(aq)}$$

(B can be a molecule or an anion).

The expression for the equilibrium constant is

$$K_{eq} = \dfrac{[BH^+][OH^-]}{[B][H_2O]}$$

and the base ionization constant, $K_b$, for the weak base is

$$K_b = K_{eq}[H_2O] = \dfrac{[BH^+][OH^-]}{[B]}$$

When $K_b \geq 1$, the base is classified as strong.

When $K_b < 1$, the base is classified as weak (Table 14-6).

## 14.7   Acid and Base Anhydrides

An acid anhydride is a nonmetallic oxide that reacts with water to form an acid.

The nonmetal in the anhydride has the same oxidation number as the nonmetal in the acid that it forms.

A basic anhydride is a metallic oxide that reacts with water to form a base.

The metal ion in the anhydride has the same charge as the metal ion in the base that it forms.

## 14.8 Acid-Base Titrations

Titration is a laboratory procedure that measures the volume of a standard solution (known concentration) required for complete reaction with another solution of unknown concentration (Figure 14-3).

To find the unknown molarity of the acid (or base), find the moles of the acid (or base) using solution stoichiometry based on the balanced chemical equation for the acid-base reaction.  (The mole map in Figure 12-11 can be used).

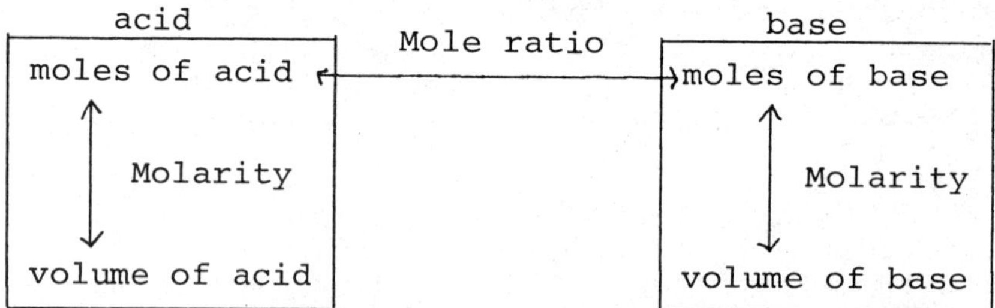


The molarity of the acid (or base) can then be found using the calculated moles and the known volume.

$$M = \frac{\text{moles acid (or base)}}{\text{L solution}}$$

## 14.9 Normality (Optional)

An equivalent of acid is the weight of acid that neutralizes one mole of hydroxide ions.

An equivalent of base is the weight of base that neutralizes one mole of hydronium ions.

One mole of a monoprotic acid contains one equivalent of acid.  One mole of a diprotic acid contains two equivalents of acid, and so on (Table 14-7).

Bases containing one hydroxide ion per formula unit have one equivalent of base for each mole of base.  One mole of base with two hydroxide ions per formula unit contains two equivalents of base.

A one normal solution, 1N, contains one equivalent of substance in one liter of solution.  Normality is a concentration defined as the number of equivalents per liter of solution,

$$N = \frac{\text{equivalents solute}}{\text{liter of solution}}$$

Normality can be calculated from molarity or determined by titration.

In acid-base reactions, one equivalent of acid neutralizes one equivalent of base:

$$H_3O^+_{(aq)} + OH^-_{(aq)} \longrightarrow 2\ H_2O_{(l)}$$

Since $eq_{acid} = eq_{base}$,

$$V_a N_a = V_b N_b$$

## 14.10   Salts

In an acid-base reaction, an ionic compound, called a salt, is always formed.  The salt is composed of cations from the base combined with anions from the acid.

An ionic equation can be written for reactions that occur in aqueous solution.  Ionic equations contain formulas for ions and molecules as they actually exist in a reaction mixture.

Those substances that exist as ions in solution are soluble ionic compounds and strong acids and bases.  These ions are written as hydrated ions in the ionic equation.

Those substances that exist as molecules in solution are water soluble molecular compounds and weak acids and bases. These substances are written in the ionic equation as molecules.

A net ionic equation contains only the formulas of the molecules or ions that actually participate in the reaction. Spectator ions are eliminated.

## 14.11   Solubilities of Salts and Other Ionic Compounds

An insoluble ionic compound, formed when two aqueous solutions of soluble ionic compounds are mixed, is called a precipitate.

General water solubility guidelines:

1. Compounds containing Group IA cations, $NH_4^+$, $NO_3^-$, $ClO_3^-$, $ClO_4^-$, or $C_2H_3O_2^-$ are usually soluble.
2. Binary compounds containing halogen anions are usually soluble, except for halides of $Ag^+$, $Hg_2^{2+}$, and $Pb^{2+}$.
3. Compounds containing $SO_4^{2-}$ are usually soluble, except when combined with $Ca^{2+}$, $Sr^{2+}$, $Ba^{2+}$, $Hg_2^{2+}$, $Hg^+$, $Pb^{2+}$, and $Ag^+$.
4. The oxides and hydroxides of Group IA elements, $Ba^{2+}$, $Sr^{2+}$, and $NH_4^+$ are soluble.
5. Most other ionic compounds are insoluble.

14.12   <u>Solubility Product Constants</u> (optional)

Slightly soluble ionic compounds exist in water as an equilibrium mixture of dissolved ions and undissolved solid.

$$M_yX_{z\,(s)} \rightleftharpoons yM^{z+}_{(aq)} + zX^{y-}_{(aq)}$$

The expression for the equilibrium constant is

$$K_{eq} = \frac{[M^{z+}]^Y[XY^-]^Z}{[M_yX_z]}$$

The concentration of $M_yX_{z\,(s)}$ corresponds to the concentration of a pure substance (since the compound is only slightly soluble) and thus is a constant.

$$K_{sp} = [K_{eq}][M_yX_z] = [M^{z+}]^Y[X^y{}^-]^Z$$

$K_{sp}$ is the solubility product constant. $K_{sp}$ values are small due to the low concentration of ions in solution. (Table 14-8).

For formula units containing the same number of ions, the higher the $K_{sp}$, the greater the solubility.  If the product of the ion concentrations (raised to their appropriate powers) exceeds the $K_{sp}$ value, a precipitate forms.

Increasing the concentration of one ion of a slightly soluble compound causes the saturation equilibrium to shift toward formation of precipitate.  This is called the common ion effect.

---

SOLUTIONS TO SELECTED TEXT STUDY QUESTIONS AND PROBLEMS
and
PRACTICE PROBLEMS

---

10.   What is the $[H_3O^+]$ in a 6.0 M solution of hydrochloric acid, $HCl_{(aq)}$?

HCl is a strong acid, which means that it ionizes completely.  To indicate this, the reaction is written with a single arrow.

$$HCl_{(aq)} + H_2O_{(1)} \longrightarrow H_3O^+_{(aq)} + Cl^-_{(aq)}$$

This reaction shows that 1 mole of HCl ionizes completely to form 1 mole of $H_3O^+_{(aq)}$ and 1 mole of $Cl^-_{(aq)}$.  (No molecular HCl remains.)  In this problem, then, 6.0 moles HCl in one liter of solution (6.0 M) will ionize to form 6.0 moles $H_3O^+$ and 6.0 moles $Cl^-$.  The concentration of $[H_3O^+]$ is 6.0 M.

12.   Hydrogen cyanide, a toxic gas, dissolves in water to form hydrocyanic acid, a weak acid:

$$HCN_{(aq)} + H_2O_{(1)} \rightleftharpoons H_3O^+_{(aq)} + CN^-_{(aq)}$$

If the following equilibrium concentrations exist,

$$[HCN] = 0.20 \ M$$
$$[H_3O^+] = 9.9 \ X \ 10^{-6} \ M$$
$$[CN^-] = 9.9 \ X \ 10^{-6} \ M$$

calculate $K_a$ for hydrocyanic acid.

Weak acids and bases ionize only slightly in aqueous solution. These reactions must be written with a double arrow.

$$HCN_{(aq)} + H_2O_{(l)} \rightleftharpoons H_3O^+_{(aq)} + CN^-_{(aq)}$$

The double arrow indicates that molecular HCN is in equilibrium with $H_3O^+_{(aq)}$ and $CN^-_{(aq)}$ ions. All three species (along with water molecules) exist in solution.

Before solving the problem, notice that molecules and ions exist for hydrocyanic acid at equilibrium. Also note that $[H_3O^+] = [CN^-]$ and that these concentrations are very small compared to [HCN]. This information is all consistent for weak acids.

$$K_a = \frac{[H_3O^+][CN^-]}{[HCN]}$$

$$= \frac{(9.9 \ X \ 10^{-6})(9.9 \ X \ 10^{-6})}{0.20}$$

$$= 4.9 \ X \ 10^{-10}$$

(Units are not included in $K_a$, $K_b$ problems.)

The small value for $K_a$ confirms that HCN is a weak acid.

***Practice Problem A

   a.  What is the $[H_3O^+]$ in a 0.30 M solution of $HNO_3$?

   b.  What is the $[H_3O^+]$ in an acetic acid solution, $HC_2H_3O_2$, that contains 0.30 M $HC_2H_3O_2$ at equilibrium. $K_a = 1.8 \ X \ 10^{-5}$

29.  A 25.00 mL sample of $H_2SO_4$ required 32.15 mL of 0.6000 M NaOH to reach the end point of a titration. What was the molarity of the $H_2SO_4$?

Write a balanced chemical equation and label it.

$$H_2SO_4 + 2\,NaOH \longrightarrow H_2O + Na_2SO_4$$

$$\begin{array}{cc} 25.00\ mL & 32.15\ mL \\ ?\ M & 0.6000\ M \end{array}$$

Calculate moles $H_2SO_4$ (use the mole map in section 14.8 as a guide).

32.15 mL NaOH $\longrightarrow$ ? L NaOH $\longrightarrow$ ? mol NaOH $\longrightarrow$ ? mol $H_2SO_4$

$$32.15\ mL\ NaOH \times \frac{10^{-3}\ L\ NaOH}{1\ mL\ NaOH} \times \frac{0.6000\ mol\ NaOH}{1\ L\ NaOH} \times \frac{1\ mol\ H_2SO_4}{2\ mol\ NaOH} =$$

$$9.645 \times 10^{-3}\ mol\ H_2SO_4$$

Calculate M $H_2SO_4$.

$$M = \frac{mol\ H_2SO_4}{L\ solution} = \frac{9.645 \times 10^{-3}\ mol\ H_2SO_4}{0.02500\ L\ solution}$$

$$= 3.858 \times 10^{-1}\ M\ H_2SO_4$$

***Practice Problem B

A 25.00 mL sample of $Ba(OH)_2$ required 45.63 mL of 0.0988 M HCl to reach the end point of a titration. What was the molarity of the $Ba(OH)_2$.

35.  Calculate the normality of a solution containing 392.4g $H_2SO_4$ in 2.00 L solution.

First, calculate the number of equivalents of $H_2SO_4$ in 392.4g. This calculation is similar to a g $\longrightarrow$ mol conversion with an additional step, mol $\longrightarrow$ eg.

$$392.4g\ H_2SO_4 \times \frac{1\ mol\ H_2SO_4}{98.08g\ H_2SO_4} \times \frac{?\ eq\ H_2SO_4}{?\ mol\ H_2SO_4} =$$

Since 1 mol $H_2SO_4$ neutralizes 2 mol $OH^-$
$(H_2SO_4 + 2\,NaOH \longrightarrow 2\,H_2O + Na_2SO_4)$;
1 mol $H_2SO_4$ = 2 eq $H_2SO_4$.

$$392.4g\ H_2SO_4 \times \frac{1\ mol\ H_2SO_4}{98.08g\ H_2SO_4} \times \frac{2\ eq\ H_2SO_4}{1\ mol\ H_2SO_4} = 8.002\ eq\ H_2SO_4$$

Calculate normality.

$$N = \frac{\text{equivalents solute}}{\text{L of solution}}$$

$$= \frac{8.002 \text{ eq } H_2SO_4}{2.00 \text{ L solu}} = 4.00 \frac{\text{eq } H_2SO_4}{L}$$

$$= 4.00 \text{ N } H_2SO_4$$

40.   What is the normality of a 2.0 M solution of $H_3PO_4$?

Normality and molarity can be interconverted using a conversion factor that equates moles to the number of equivalents. In $H_3PO_4$ solutions,

$$1 \text{ mol } H_3PO_4 = 3 \text{ eq } H_3PO_4$$

Write the given concentration using a numerator and denominator and multiply it by the mol $\longrightarrow$ eq conversion factor.

$$\frac{2.0 \text{ mol } H_3PO_4}{\text{L solution}} \times \frac{3 \text{ eq } H_3PO_4}{1 \text{ mol } H_3PO_4} = \frac{6.0 \text{ eq } H_3PO_4}{\text{L solution}}$$

$$= 6.0 \text{ N } H_3PO_4$$

44.   Calculate the volume of 0.4000 N base required to titrate 15.00 mL of 1.000 N $HNO_3$.

At the end point of a titration,

$$eq_{acid} = eq_{base}$$

Since, $N = \dfrac{\text{equivalents of solute}}{\text{volume solution (L)}}$

$$eq = N \times V$$

Thus,

$$N_A \times V_A = N_B \times V_B$$

State the problem.

$$N_A = 1.000 \text{ N } HNO_3 \qquad N_B = 0.4000 \text{ N base}$$
$$V_A = 15.00 \text{ mL} \qquad V_B = ?$$
$$V_B = V_A\left(\frac{N_A}{N_B}\right)$$
$$= 15.00 \text{ mL}\left(\frac{1.000 \text{ N}}{0.4000 \text{ N}}\right) = 37.50 \text{ mL base}$$

***Practice Problem C

What volume of 0.200 N $H_3PO_4$ is required to neutralize 3.80g of $Ca(OH)_2$?

50.  Write the balanced nonionic equation for any reactions that occur when aqueous solutions of the following are mixed.

a.  $AlCl_3$ and NaOH

The problem states that the reactants are aqueous solutions, so they exist in solution as ions.  For a reaction to occur an insoluble ionic compound or molecular compound must be formed.

Begin by writing a balanced chemical equation.

$$AlCl_3 + 3\ NaOH \longrightarrow 3\ NaCl + Al(OH)_3$$

Examine the products.  Rule 4 of the solubility guidelines in section 14.11 indicates that $Al(OH)_3$ is insoluble.  A reaction occurs.

b.  $Ba(C_2H_3O_2)_2$ and $K_2SO_4$

$$Ba(C_2H_3O_2)_2 + K_2\ SO_4 \longrightarrow BaSO_4 + 2\ KC_2H_3O_2$$

Barium sulfate is an insoluble ionic compound, so this reaction occurs.

c.  $AgNO_3$ and $NaClO_3$

$$AgNO_3 + NaClO_3 \longrightarrow AgClO_3 + NaNO_3$$

$NaNO_3$ is soluble.  The solubility rules indicate that many $ClO_3^-$ compounds are soluble. However, $AgClO_3$ is insoluble. (Many $Ag^+$ compounds are insoluble.)

d.  $FeCl_2$ and LiOH

$$FeCl_2 + 2\ NaOH \longrightarrow Fe\ (OH)_2 + 2\ NaCl$$

$Fe(OH)_2$ is an insoluble ionic compound, so this reaction occurs.

51.  Write the total ionic compound for any reaction that occurs in question 50.

a.  Write soluble ionic compounds as ions.  Write unsoluble ionic compounds as the formula unit with a subscript (s) or ↓.

$$Al^{3+}_{(aq)} + 3\ Cl^-_{(aq)} + 3\ Na^+_{(aq)} + 3\ OH^-_{(aq)} \longrightarrow 3\ Na^+_{(aq)} + 3\ Cl^-_{(aq)} + Al(OH)_{3\,(s)}$$

Notice that the subscripts in ionic formula units become coefficients in the ionic equation (except for polyatomic ions, see part b).

b.  $$Ba^{2+}_{(aq)} + 2\ C_2H_3O_2^-{}_{(aq)} + 2\ K^+_{(aq)} + SO_4^{2-}{}_{(aq)} \longrightarrow BaSO_{4\,(s)} + 2\ K^+_{(aq)} + C_2H_3O_2^-{}_{(aq)}$$

The subscript 2 for the parenthesis around the acetate ion becomes the coefficient in the ionic equation.

c.  $$Ag^+_{(aq)} + NO_3^-{}_{(aq)} + Na^+_{(aq)} + ClO_3^-{}_{(aq)} \longrightarrow AgClO_{3\,(s)} + Na^+_{(aq)} + NO_3^-{}_{(aq)}$$

d.  $$Fe^{2+}_{(aq)} + 2\ Cl^-_{(aq)} + 2\ Na^+_{(aq)} + 2\ OH^-_{(aq)} \longrightarrow Fe(OH)_{2\,(s)} + 2\ Na^+_{(aq)} + 2\ Cl^-_{(aq)}$$

52.  Write the net ionic equation for any reaction that would occur in question 50.

a.  $$Al^{3+}_{(aq)} + 3\ Cl^-_{(aq)} + 3\ Na^+_{(aq)} + 3\ OH^-_{(aq)} \longrightarrow 3\ Na^+_{(aq)} + 3\ Cl^-_{(aq)} + Al(OH)_{3\,(s)}$$

Cancel out the spectator ions in the ionic equation above.

$$Al^{3+}_{(aq)} + 3\ \cancel{Cl^-}_{(aq)} + 3\ \cancel{Na^+}_{(aq)} + 3\ OH^-_{(aq)} \longrightarrow 3\ \cancel{Na^+}_{(aq)} + 3\ \cancel{Cl^-}_{(aq)} + Al(OH)_{3\,(s)}$$

Net ionic equation:  $$Al^{3+}_{(aq)} + 3\ OH^-_{(aq)} \longrightarrow Al(OH)_{3\,(s)}$$

b.  $$Ba^{2+}_{(aq)} + 2\ \cancel{C_2H_3O_2^-}{}_{(aq)} + 2\ \cancel{K^+}_{(aq)} + SO_4^{2-}{}_{(aq)} \longrightarrow BaSO_{4\,(s)} + 2\ \cancel{K^+}_{(aq)} + \cancel{C_2H_3O_2^-}{}_{(aq)}$$

Net ionic equation:  $$Ba^{2+}_{(aq)} + SO_4^{2-}{}_{(aq)} \longrightarrow BaSO_{4\,(s)}$$

c.  $$Ag^+_{(aq)} + \cancel{NO_3^-}{}_{(aq)} + \cancel{Na^+}_{(aq)} + ClO_3^-{}_{(aq)} \longrightarrow AgClO_{3\,(s)} + \cancel{Na^+}_{(aq)} + \cancel{NO_3^-}{}_{(aq)}$$

Net ionic equation:  $$Ag^+_{(aq)} + ClO_3^-{}_{(aq)} \longrightarrow AgClO_{3\,(s)}$$

d.  $$Fe^{2+}_{(aq)} + 2\ \cancel{Cl^-}_{(aq)} + 2\ \cancel{Na^+}_{(aq)} + 2\ OH^-_{(aq)} \longrightarrow Fe(OH)_{2\,(s)} + 2\ \cancel{Na^+}_{(aq)} + 2\ \cancel{Cl^-}_{(aq)}$$

Net ionic equation:  $$Fe^{2+}_{(aq)} + 2\ OH^-_{(aq)} \longrightarrow Fe(OH)_{2\,(s)}$$

***Practice Problem D

Write ionic equations and net ionic equations for each of the following reactions occurring in aqueous solution.

a.  $Pb(NO_3)_2 + Na_2SO_4$

b.  $H_2SO_4 + NaOH$

c.  $LiCl + KOH$

54.  The solubility of $Cd(OH)_2$ at 25°C is $1.7 \times 10^{-5}$ mol/L. Calculate the $K_{sp}$ of $Cd(OH)_2$ at 25°C.

$Cd(OH)_2$, which is slightly soluble in water, exists in water as an equilibrium mixture of dissolved ions and undissolved solid.

Write the equation for this equilibrium.

$$Cd(OH)_2 \rightleftharpoons Cd^{2+}_{(aq)} + 2\ OH^-_{(aq)}$$

The $K_{sp}$ expression is:

$$K_{sp} = [Cd^{2+}][OH^-]^2$$

The solubility of $Cd(OH)_2$ is given as $1.7 \times 10^{-5}$ mol/L. This means that $1.7 \times 10^{-5}$ mol of $Cd(OH)_2$ will dissolve (form ions) in 1 liter of solution. When 1 mol of $Cd(OH)_2$ dissolves, 1 mol $Cd^{2+}$ ions and 2 mol $OH^-$ ions go into solution.

In 1 L solution, $1.7 \times 10^{-5}$ mol $Cd(OH)_2$ dissolves to form:

$[Cd^{2+}] = 1.7 \times 10^{-5}$ mol/L $Cd^{2+}$

$[OH^-] = 2(1.7 \times 10^{-5}$ mol/L $OH^-) = 3.4 \times 10^{-5}$ mol/L $OH^-$

Substitute these values into the Ksp expression above.

$$Ksp = (1.7 \times 10^{-5})(3.4 \times 10^{-5})^2$$

$$Ksp = 2.0 \times 10^{-14}$$

69.  At 25°C, the Ksp of $CaF_2$ is $4.0 \times 10^{-11}$. What is the solubility of $CaF_2$ in mol/L?

Write the equation to describe the equilibrium.

$$CaF_{2(s)} \rightleftharpoons Ca^{2+}_{(aq)} + 2\ F^-_{(aq)}$$

When 1 mol $CaF_2$ dissolves, 1 mol $Ca^{2+}$ ions form.  Therefore, the molarity of $CaF_2$ dissolves is equal to the molarity of $Ca^{2+}$ in solution, $[Ca^{2+}]$.

Write the Ksp expression.

$$K_{sp} = [Ca^{2+}][F^-]^2$$

Assume that x moles of $CaF_2$ dissolves in 1 L of solution.  The concentration of ions formed is:

$$[Ca^{2+}] = x \text{ mol/L}$$

$$[F^-] = 2x \text{ mol/L}$$

Substitute these values into the Ksp expression above.

$$K_{sp} = (x)(2x)^2$$
$$K_{sp} = 4x^3$$
$$4.0 \times 10^{-11} = 4x^3$$
$$1.0 \times 10^{-11} = x^3$$
$$\sqrt[3]{1.0 \times 10^{-11}} = x$$

There are several methods for finding cube roots on the calculator.  One method involves writing the root as a fractional exponent.  This exponent can then be written as a decimal.

$$(1.0 \times 10^{-11})^{1/3} = x$$

$$(1.0 \times 10^{-11})^{.33\overline{3}} = x$$

Enter 1.0 $\boxed{EE}$ 11 $\boxed{+/-}$ $\boxed{y^x}$ .333333 $\boxed{=}$ .

The display reads $2.1544 \times 10^{-4}$.  The answer is rounded off to 2 significant figures.

$$x = 2.15 \times 10^{-4}$$

Since the assumption was made that x moles of $CaF_2$ dissolves in 1 L of solution, the solubility of $CaF_2$ is $2.15 \times 10^{-4}$ mol/L.

***Practice Problem E

    a.  The solubility of silver carbonate, $Ag_2CO_3$, is $1.27 \times 10^{-4}$ mol/L.  Calculate the value of $K_{sp}$ for silver carbonate.

b.  The $K_{sp}$ of silver chromate, $Ag_2CrO_4$, is $1.1 \times 10^{-12}$. Compare the $K_{sp}$ for $Ag_2CrO_4$ to the $K_{sp}$ for $Ag_2CO_3$ calculated above.  Which silver compound is more soluble?

c.  Calculate the solubility of $Ag_2CrO_4$.  Compare the solubility of $Ag_2CrO_4$ to the solubility of $Ag_2CO_3$ given in part a.  (The results should confirm that the compound selected as more soluble in part b, is indeed more soluble.)

---

## SELF-TEST

Circle the correct answer in the following multiple choice questions.

1.  The conjugate base of ammonia, $NH_3$, is

    a.  $NH_2^-$
    b.  $NH_4^+$
    c.  $H_3O^+$
    d.  $OH^-$

2.  In the following reaction,

$$HS^- + H_2O \rightleftharpoons H_3O^+ + S^{2-}$$

one of the conjugate acid-base pairs is

    a.  $HS^-$ and $H_2O$
    b.  $HS^-$ and $H_3O^+$
    c.  $HS^-$ and $S^{2-}$
    d.  $H_3O^+$ and $S^{2-}$

3.   In solution, hydrofluoric acid, HF, exists as

a.   HF molecules only
b.   $H_3O^+$ ions only
c.   $H_3O^+$ and $F^-$ ions only
d.   $H_3O^+$, $F^-$ ions, and HF molecules

4.   Which of the following acids is the weakest?

a.   $HC_2H_3O_2$, Ka = 1.8 X $10^{-5}$

b.   HCN,      Ka = 4.9 X $10^{-10}$

c.   $HNO_2$,     Ka = 4.5 X $10^{-4}$

d.   $H_2CO_3$,    Ka = 4.3 X $10^{-7}$

5.   The compound that forms when the anhydride, $Cl_2O_3$, combines
with water is

a.   HClO
b.   $HClO_2$
c.   $HClO_3$
d.   $HClO_4$

6.   The equivalent weight of phosphorous acid, $H_3PO_3$ is

a.   41.0g
b.   246g
c.   27.3g
d.   82.0g

7.   10.00g of $Sr(OH)_2$, dissolved to make 250.0 mL of solution,
has a normality of

a.   0.658 N
b.   0.329 N
c.   0.164 N
d.   2.00 N

8.   Which of the following is a correct net ionic equation for
the reaction,

$$K_2SO_4 + 2\ AgNO_3 \longrightarrow Ag_2SO_4 + 2\ KNO_3$$

a.   $2\ K^+_{(aq)} + 2\ NO_3^-{}_{(aq)} \longrightarrow 2\ KNO_3{}_{(s)}$

b.   $2\ K^+_{(aq)} + SO_4^{2-}{}_{(aq)} + 2\ Ag^+_{(aq)} + 2\ NO_3^-{}_{(aq)} \longrightarrow Ag_2SO_4{}_{(s)} + 2\ K^+_{(aq)} + 2\ NO_3^-{}_{(aq)}$

c.   $2\ Ag^+_{(aq)} + SO_4^{2-}{}_{(aq)} \longrightarrow Ag_2SO_4{}_{(s)}$

d.   No reaction, both reactants and products contain only
ions.

9.  What volume of 1.50 N $HC_2H_3O_2$ is required to neutralize 25.0 mL of 2.00 N KOH solution?

    a.  18.8 mL
    b.  12.0 mL
    c.  30.0 mL
    d.  33.3 mL

10. Calculate the $[H_3O^+]$ concentration of hypochlorous acid, HOCl, that contains 0.500 M HOCl at equilibrium. $K_a = 3.5 \times 10^{-8}$.

    a.  $1.32 \times 10^{-4}$ M
    b.  $8.57 \times 10^{-9}$ M
    c.  $2.65 \times 10^{-4}$ M
    d.  $3.50 \times 10^{-8}$ M

---

## ANSWERS TO PRACTICE PROBLEMS

**Practice Problem A**

a.  Since $HNO_3$ is a strong acid, $HNO_3$ ionizes completely.

$$HNO_{3(aq)} + H_2O_{(1)} \longrightarrow H_3O^+_{(aq)} + NO_3^-_{(aq)}$$

0.30 M $HNO_3$ will ionize to form
0.30 M $H_3O^+$ and 0.30 M $NO_3^-_{(aq)}$
(No molecular $HNO_3$ exists.)

b.  Since $HC_2H_3O_2$ is a weak acid, $HC_2H_3O_2$ ionizes only slightly.

$$HC_2H_3O_{2(aq)} + H_2O_{(1)} \rightleftharpoons H_3O^+_{(aq)} + C_2H_3O_2^-_{(aq)}$$

At equilibrium, $[HC_2H_3O_2] = 0.30$ M.  Since $K_a$ is given and $[H_3O^+] = [C_2H_3O_2^-]$ at equilibrium, $[H_3O^+]$ can be calculated.

$$K_a = \frac{[H_3O^+][C_2H_3O_2^-]}{[H_2C_2H_3O_2]}$$

Since $[H_3O^+] = [C_2H_3O_2^-]$, substitute $[H_3O^+]$ into the equation.

$$K_a = \frac{[H_3O^+][H_3O^+]}{[HC_2H_3O_2]}$$

Rearrange to solve for $[H_3O^+]$.

$$[H_3O^+]^2 = K_a[HC_2H_3O_2]$$
$$= 1.8 \times 10^{-5}(0.30) = 5.4 \times 10^{-6}$$
$$[H_3O^+] = \sqrt{5.4 \times 10^{-6}} = 2.3 \times 10^{-3} \text{ M}$$

$[H_3O^+]$ is much smaller than $[HC_2H_3O_2]$ at equilibrium.

The following comparisons should be noted for part a (strong acid) and part b (weak acid).

1. $[H_3O^+]$ for the strong acid is much larger than $[H_3O^+]$ for the weak acid.
2. No calculation is required for a strong acid. $[H_3O^+]$ can be found directly from the concentration of the acid.
3. A strong acid is composed of ions while a weak acid is composed of molecules and ions.

## Practice Problem B

Write a balanced chemical equation and label it.

$$Ba(OH)_2 + 2\ HCl \longrightarrow BaCl_2 + 2\ H_2O$$

$$\begin{array}{cc} 25.00\ mL & 45.63\ mL \\ ?\ M & 0.0988M \end{array}$$

Calculate moles $Ba(OH)_2$.

$$45.63\ mL\ HCl \longrightarrow ?\ L\ HCl \longrightarrow ?\ mol\ HCl \longrightarrow ?\ mol\ Ba(OH)_2$$

$$45.63\ \cancel{mL\ HCl} \times \frac{10^{-3}\ \cancel{L\ HCl}}{1\ \cancel{mL\ HCl}} \times \frac{0.0988\ \cancel{mol\ HCl}}{1\ \cancel{L\ HCl}} \times \frac{1\ mol\ Ba(OH)_2}{2\ \cancel{mol\ HCl}} =$$

$$2.254 \times 10^{-3}\ mol\ Ba(OH)_2$$

Calculate $M\ Ba(OH)_2$.

$$M = \frac{mol\ Ba(OH)_2}{L\ solution} = \frac{2.254 \times 10^{-3}\ mol\ Ba(OH)_2}{0.02500\ L\ solution}$$

$$= 9.016 \times 10^{-2}\ M\ Ba(OH)_2$$

## Practice Problem C

Calculate the number of equivalents of $Ca(OH)_2$.

$$3.80g\ \cancel{Ca(OH)_2} \times \frac{1\ \cancel{mol\ Ca(OH)_2}}{74.1g\ \cancel{Ca(OH)_2}} \times \frac{2\ eq\ Ca(OH)_2}{1\ \cancel{mol\ Ca(OH)_2}} = 0.103\ eq\ Ca(OH)_2$$

At the end point of the titration,

$$eq_{H_3PO_4} = eq_{Ca(OH)_2} = 0.103\ eq$$

Calculate the volume of $H_3PO_4$, using normality $\left(0.200\ N = \dfrac{0.200\ eq}{1\ L}\right)$ as a conversion factor.

$$0.103\ \cancel{eq\ H_3PO_4} \times \frac{1\ L\ solution}{0.200\ \cancel{eq\ H_3PO_4}} = 0.515\ L\ H_3PO_4\ solution$$

$$= 515\ mL\ H_3PO_4\ solution$$

## Practice Problem D

a. Begin by writing a balanced chemical equation.

$$Pb(NO_3)_2 + Na_2SO_4 \longrightarrow 2\ NaNO_3 + PbSO_4$$

Check the solubility guidelines in section 14.11. $PbSO_4$ is insoluble.

Next, write the total ionic equation.

$$Pb^{2+}_{(aq)} + 2\ NO_3{}^-_{(aq)} + 2\ Na^+_{(aq)} + SO_4{}^{2-}_{(aq)} \longrightarrow 2\ Na^+_{(aq)} + 2\ NO_3{}^-_{(aq)} + PbSO_4{}_{(s)}$$

Cancel out the spectator ions.

$$Pb^{2+}_{(aq)} + \cancel{2\ NO_3{}^-_{(aq)}} + \cancel{2\ Na^+_{(aq)}} + SO_4{}^{2-}_{(aq)} \longrightarrow \cancel{2\ Na^+_{(aq)}} + \cancel{2\ NO_3{}^-_{(aq)}} + PbSO_4{}_{(s)}$$

The net ionic equation:

$$Pb^{2+}_{(aq)} + SO_4{}^{2-}_{(aq)} \longrightarrow PbSO_4{}_{(s)}$$

b.  Begin by writing a balanced chemical equation.

$$H_2SO_4 + 2\ NaOH \longrightarrow Na_2SO_4 + 2\ H_2O$$

NaOH and $Na_2SO_4$ are soluble ionic compounds and $H_2SO_4$ is a strong acid.  These three substances exist in aqueous solutions as ions.  $H_2O$ is a molecular compound and it exists in solution as a molecule.  The subscripts of molecules identify the state of matter, so water is written $H_2O_{(l)}$.

The total ionic equation is

$$2\ H^+_{(aq)} + SO_4{}^{2-}_{(aq)} + 2\ Na^+_{(aq)} + 2\ OH^-_{(aq)} \longrightarrow 2\ Na^+_{(aq)} + SO_4{}^{2-}_{(aq)} + 2\ H_2O_{(l)}$$

Cancel out the spectator ions.

$$2\ H^+_{(aq)} + \cancel{SO_4{}^{2-}_{(aq)}} + \cancel{2\ Na^+_{(aq)}} + 2\ OH^-_{(aq)} \longrightarrow \cancel{2\ Na^+_{(aq)}} + \cancel{SO_4{}^{2-}_{(aq)}} + 2\ H_2O_{(l)}$$

Net ionic equation:

$$2\ H^+_{(aq)} + 2\ OH^-_{(aq)} \longrightarrow 2\ H_2O_{(l)}$$

This reaction is typical of acid-base neutralization reactions.  The formation of water molecules causes this reaction to occur.

c.  Begin by writing a balanced chemical equation.

$$LiCl + KOH \longrightarrow LiOH + KCl$$

The solubility guidelines indicate all four substances are soluble.  The reactants exist as ions and the products exist as ions.  No net reaction occurs.

$$\cancel{Li^+_{(aq)}} + \cancel{Cl^-_{(aq)}} + \cancel{K^+_{(aq)}} + \cancel{OH^-_{(aq)}} \longrightarrow \cancel{Li^+_{(aq)}} + \cancel{OH^-_{(aq)}} + \cancel{K^+_{(aq)}} + \cancel{Cl^-_{(aq)}}$$

Practice Problem E

a.  The equation for the equilibrium is:

$$Ag_2CO_{3(s)} \rightleftharpoons 2\,Ag^+_{(aq)} + CO_3{}^{2-}{}_{(aq)}$$

The $K_{sp}$ expression is:

$$K_{sp} = [Ag^+]^2[CO_3{}^{2-}]$$

In 1 liter of solution $1.27 \times 10^{-4}$ mol $Ag_2CO_3$ dissolves to form:

$$[Ag^+] = 2(1.27 \times 10^{-4})\,mol/L\ Ag^+ = 2.54 \times 10^{-4}\,mol/L\ Ag^+$$
$$[CO_3{}^{2-}] = 1.27 \times 10^{-4}\,mol/L\ CO_3{}^{2-}$$

Substitute these values into the Ksp expression above.

$$K_{sp} = (2.54 \times 10^{-4})^2(1.27 \times 10^{-4})$$
$$K_{sp} = 8.2 \times 10^{-12}$$

b.  The $K_{sp}$ for $Ag_2CrO_4$ is smaller than the $K_{sp}$ for $Ag_2CO_3$. Both formula units contain the same number of ions, so $Ag_2CO_3$ has a greater solubility than $Ag_2CrO_4$.

c.  The equation for the equilibrium is:

$$Ag_2CrO_{4(s)} \rightleftharpoons 2\,Ag^+_{(aq)} + CrO_4{}^{2-}{}_{(aq)}$$

The $K_{sp}$ expression is:

$$K_{sp} = [Ag^+]^2[CrO_4{}^{2-}]$$

Assume that x moles of $Ag_2CrO_4$ dissolves in 1 L of solution.  The concentration of ions formed is:

$$[Ag^+] = 2x\ mol/L$$
$$[CrO_4{}^{2-}] = x\ mol/L$$

Substitute these values into the Ksp expression above.

$$K_{sp} = (2x)^2(x)$$
$$K_{sp} = 4x^3$$
$$1.1 \times 10^{-12} = 4x^3$$
$$2.8 \times 10^{-13} = x^3$$
$$\sqrt[3]{2.8 \times 10^{-13}} = x$$
$$(2.8 \times 10^{-13})^{.333\overline{3}} = x$$
$$6.5 \times 10^{-5} = x$$

The solubility of $Ag_2CrO_4$ is $6.5 \times 10^{-5}$ mol/L.

The solubility of $Ag_2CO_3$ given in part a, $1.27 \times 10^{-4}$ mol/L is larger than the calculated solubility of $Ag_2CrO_4$ in part c.  This confirmed the prediction made in part b.

---

## ANSWERS TO SELF-TEST

1. a
2. c
3. d
4. b
5. b
6. c
7. a
8. c
9. d
10. a

---

## SUMMARY OUTLINE

15.1    <u>Electrolytes</u>

Electrolytes are substances that conduct electricity when they are melted or dissolved in water (Table 15-1).

Strong electrolytes are compounds that dissociate completely into ions when dissolved in water.  Strong acids, strong bases and soluble ionic compounds are strong electrolytes.  Aqueous solutions of strong electrolytes have high electrical conductivity.

Weak electrolytes are compounds that only partially dissociate into ions when dissolved in water.  Weak acids, weak bases and slightly soluble ionic compounds are weak electrolytes.  Aqueous solutions of weak electrolytes have low electrical conductivity.

Nonelectrolytes have no ions in solution.  Molecular compounds (other than acids) and insoluble ionic compounds are nonelectrolytes.  Aqueous solutions of nonelectrolytes have no electrical conductivity (Table 15-1).

15.2.    <u>The Ionization of Water</u>

Ionization of water occurs according to the following equation:

$$H_2O_{(l)} + H_2O_{(l)} \rightleftharpoons H_3O^+_{(aq)} + OH^-_{(aq)}$$

Water is a weak acid,

$$\underset{\text{acid}_1}{H_2O_{(l)}} + H_2O_{(l)} \rightleftharpoons \underset{\text{base}_2}{H_3O^+ + OH^-}$$

and water is a weak base.

$$\underset{\text{base}_1}{H_2O_{(l)} + H_2O_{(l)}} \rightleftharpoons \underset{\text{acid}_2}{H_3O^+ + OH^-}$$

The expression for the equilibrium constant of water is derived below:

$$K_{eq} = \frac{[H_3O^+][OH^-]}{[H_2O]^2}$$

The $[H_2O]$ is constant and $K_{eq}$ is constant.  Thus,

$$K_{eq}[H_2O]^2 = [H_3O^+][OH^-]$$

$$K_w = [H_3O^+][OH^-]$$

where $K_w$, is the ionization constant of water, called the ion-product constant.  At 25°C, $K_w = 1.00 \times 10^{-14}$.  The small $K_w$ value reflects the very small quantity of ions present in pure water.

The product, $[H_3O^+][OH]^-$ must equal $1.00 \times 10^{-14}$.  If acid is added to water, $[H_3O^+]$ increases and $[OH^-]$ must decrease.  If base is added to water, $[OH^-]$ increases and $[H_3O^+]$ must decrease.

When $[H_3O^+] > [OH^-]$, the solution is acidic.
$[H_3O^+] < [OH^-]$, the solution is basic.
$[H_3O^+] = [OH^-]$, the solution is neutral.

## 15.3 <u>pH</u>

pH is defined as the negative logarithm of $[H_3O^+]$.

$$pH = -\log [H_3O^+]$$

A neutral solution has $[H_3O^+] = 1.00 \times 10^{-7}$; pH = 7.

An acidic solution has $[H_3O^+] > 1.00 \times 10^{-7}$; pH < 7.

A basic solution has $[H_3O^+] < 1.00 \times 10^{-7}$; pH > 7.

pOH is defined as the negative logarithm of $[OH^-]$.

$$pOH = -\log [OH^-]$$

Thus,

$$pH + pOH = 14 \text{ (Table 15-2)}.$$

## 15.4 <u>pH Scale</u>

The pH scale expresses levels of acidity in aqueous solutions. The pH values usually range from 0 to 14.

15.5   <u>Measurement of pH</u>

Acid-base indicators can be used to estimate the pH of a solution.  An indicator is a weak acid or base.

The ionization reaction for a general indicator, HInd is

$$HInd_{(aq)} + H_2O_{(l)} \rightleftharpoons H_3O^+_{(aq)} + Ind^-_{(aq)}$$

HInd and Ind$^-$ forms have different colors (Table 15-3).

pH can be measured precisely with a pH meter, an instrument that measures hydronium ions in solution.

15.6   <u>Additional pH Calculations</u> (optional)

When $[H_3O^+] = 1 \times 10^n$ M (n is zero or a negative whole number),

$$pH = -n$$

For other values of $[H_3O^+]$, pH will be a decimal number.

Antilogs are used to find $[H_3O^+]$ when pH is given.

When a weak acid, HA, is ionized according to the reaction,

$$HA_{(aq)} + H_2O_{(l)} \rightleftharpoons H_3O^+_{(aq)} + A^-_{(aq)}$$

$$[HA]_{dissociated} = [H_3O^+]\ acid$$
$$[HA]_{equilibrium} = [HA]_{initial} - [HA]_{dissociated}$$

15.7   <u>Hydrolysis</u> (optional)

Many salts contain a cation or anion that can react with water, making aqueous solutions that are either acidic or basic.  The reaction of a specific ion with water to affect the pH is called hydrolysis.

When a salt containing an anion from a weak acid is dissolved in water, the anion reacts with water according to the following reaction:

$$A^-_{(aq)} + H_2O_{(l)} \rightleftharpoons HA_{(aq)} + OH^-_{(aq)}$$
$$\text{base}_1 \qquad\qquad\qquad \text{acid}_2$$

$A^-$ is a conjugate base of the weak acid HA, making $A^-$ a strong base.  (It's a strong base because the reaction to the right readily occurs.  When $A^-$ accepts protons from water, the weak acid, HA, that forms has little tendency to ionize.  The reaction to the left is minimal.)  The hydrolysis of $A^-$ results in the formation of $OH^-$; the solution is basic (pH > 7).

When a salt containing a cation from a weak base is dissolved in water, the cation reacts with water according to the following reaction:

$$(Base)H^+_{(aq)} + H_2O_{(l)} \; \rightleftharpoons \; Base_{(aq)} + H_3O^+_{(aq)}$$

$$\underbrace{acid_1} \qquad\qquad\qquad \underbrace{base_2}$$

$(Base)H^+$ is a conjugate acid of the weak base, making $(Base)H^+$ a strong acid. (It's a strong acid because the reaction to the right readily occurs. When $(Base)H^+$ donates protons to water, the weak base that forms has little tendency to react in the direction to the left.)  The hydrolysis of $(Base)H^+$ results in the formation of $H_3O^+$; the solution is acidic (pH < 7).

Anions of strong acids and cations of strong bases do not hydrolyze in solution.

The hydrolysis of salts is summarized in the following table:

| type of salt | ion hydrolyzed | pH of solution |
|---|---|---|
| anion SA - cation SB | none | neutral |
| anion WA - cation SB | anion | basic |
| anion SA - cation WB | cation | acidic |
| anion WA - cation WB | anion, cation | varies with salt |

SA is strong acid, SB is strong base, WA is weak acid and WB is weak base.

## 15.8   Buffers

Buffer solutions are able to react with moderate amounts of both $H_3O^+$ and $OH^-$ with little change in the pH of the solution.

Buffer solutions contain either a weak acid and a salt of that weak acid or a weak base and a salt of that weak base (conjugate acid-base pairs).  The two components of a buffer solution are called a buffer pair (Table 15-4).

An HA/A buffer system resists pH changes because of the large amount of un-ionized HA,

$$HA_{(aq)} + H_2O_{(l)} \; \rightleftharpoons \; H_3O^+_{(aq)} + A^-_{(aq)}$$

and the large amount of $A^-$ in solution.

$$MA \xrightarrow{H_2O} M^+ + A^-_{(aq)}$$

($M^+$ is a metal cation.)

When moderate quantities of $H_3O^+$ ions are added, the excess $H_3O^+$ ions are removed by a shift in the equilibrium to the left.

$$HA_{(aq)} + H_2O_{(l)} \; \rightleftharpoons \; H_3O^+_{(aq)} + A^-_{(aq)}$$

$$\longleftarrow \text{ increase in } [H_3O^+]$$

Moderate amounts of added $OH^-$ are neutralized by the $H_3O^+$ ions that are present. This loss of $H_3O^+$ shifts the equilibrium to the right, making more $H_3O^+$.

$$HA_{(aq)} + H_2O_{(l)} \rightleftharpoons H_3O^+_{(aq)} + A^-_{(aq)}$$

decrease in $[H_3O^+] \longrightarrow$

---

### SOLUTIONS TO SELECTED STUDY QUESTIONS AND PROBLEMS
### and
### PRACTICE PROBLEMS

---

16.  For a 0.00100 M KOH solution,

a.  What is the $[OH^-]$?

KOH is a strong base which means it ionizes completely in water.

$$KOH_{(s)} \xrightarrow{H_2O} K^+_{(aq)} + OH^-_{(aq)}$$

This reaction shows that 1 mole of KOH ionizes in water to form 1 mole of $K^+$ ions and 1 mole of $OH^-$ ions. For the given solution, 0.00100 M KOH dissociates to form 0.00100 M $OH^-$.

$$[OH^-] = 0.00100 \text{ M} = 1.00 \times 10^{-3} \text{ M}$$

b.  What is the $[H_3O^+]$?

Since $[OH^-]$ is known, and $K_w$ is known, the expression

$$K_w = [H_3O^+][OH^-]$$

can be used to find $[H_3O^+]$.

$$[H_3O^+] = \frac{K_w}{[OH^-]}$$

$$= \frac{1.00 \times 10^{-14}}{1.00 \times 10^{-3}}$$

$$= 1.00 \times 10^{-11}$$

c.  What is the pH?

$$pH = -\log [H_3O^+]$$

$$= -\log (1.00 \times 10^{-11})$$

$$= -(-11)$$

$$= 11$$

No logarithm tables or calculators are needed to find pH when $[H_3O^+]$ is $1.00 \times 10^n$. If n is zero or a negative whole number, pH = -n.

d.    What is the pOH?

$$\text{Since pH + pOH} = 14$$
$$\text{pOH} = 14 - \text{pH}$$
$$\text{pOH} = 14 - 11$$
$$= 3$$

22.    What is the pH of each of the following solutions?

a.    $[H_3O^+] = 0.25$ M

$$\log [H_3O^+] = \log (2.5 \times 10^{-1})$$
$$= \log 2.5 + \log 10^{-1}$$

The log of 2.5 can be found using log tables (Appendix B) or a calculator.  To find the logarithm of 2.5 using the calculator,

Press 2.5 $\boxed{\log}$

The display reads 0.397940009.  The number of significant digits in the original number dictates the number of significant digits to the right of the decimal point in the logarithm.  The log 2.5 = 0.40.

$$\log [H_3O^+] = 0.40 + (-1)$$
$$= -0.602$$
$$= -0.60$$
$$\text{pH} = -\log [H_3O^+]$$
$$= -(-0.60) = 0.60$$

The log $2.5 \times 10^{-1}$ can be found using a calculator without separating the numerical and exponential part of the number, finding separate logs and then adding.  Instead,

Press 2.5 $\boxed{\text{EE}}$ 1 $\boxed{+/-}$ $\boxed{\log}$.

The display reads $-0.60206$ 00.  The answer is reported as $-0.60$ with two significant figures to the right of the decimal point.

b.    $[OH^-] = 0.35$ M

There are two different methods to solve this problem.

Method 1:

Calculate $[H_3O^+]$.

$$K_w = [H_3O^+][OH^-]$$
$$[H_3O^+] = \frac{K_w}{[OH^-]}$$
$$= \frac{1.00 \times 10^{-14}}{0.35}$$
$$= 2.9 \times 10^{-14}$$

Calculate pH.

$$\log [H_3O^+] = \log (2.9 \times 10^{-14})$$
$$= \log 2.9 + \log 10^{-14}$$
$$= 0.46 + (-14)$$
$$= -13.54.$$

The original concentration has 2 significant digits, so the logarithm is reported with two digits to the right of the decimal point.

$$pH = -\log [H_3O^+]$$
$$= - (-13.54) = 13.54$$

Method 2:

Calculate pOH.

$$\log [OH^-] = \log (3.5 \times 10^{-1})$$
$$= \log 3.5 + \log 10^{-1}$$
$$= 0.54 + (-1)$$
$$= -0.46$$
$$pOH = -\log [OH^-]$$
$$= (-0.46) = 0.46$$

Calculate pH.

$$pH + pOH = 14$$
$$pH = 14 - pOH$$
$$pH = 14 - (0.46)$$
$$= 13.54$$

Compare the two methods.  The first method relates $[OH^-]$ to $[H_3O^+]$ using the formula

$$K_w = [H_3O^+][OH^-].$$

The second uses the logarithm of that same formula to relate pOH to pH.

$$\log K_w = \log [H_3O^+] + \log [OH^-] = -14$$
$$pH + pOH = 14$$

Consult Appendix C in the text for answers to parts c and d.

***Practice Problem A

What is the pH of the following solutions?

a.   $[H_3O^+] = 0.0026$ M

b.   $[OH^-] = 0.0026$ M

c.   0.055 M $HNO_3$

d.   0.055 M $Ba(OH)_2$

24.    What is the $[H_3O^+]$ of each of the following solutions?

a.   pH = 3.05

The pH of a solution is the negative power to which 10 must be raised to describe the molar concentration of $H_3O^+$. This can be expressed in logarithms as,

$$pH = -\log [H_3O^+]$$

The equivalent exponential form is

$$[H_3O^+] = 10^{-pH} \text{ mol/L.}$$

Thus,

$$[H_3O^+] = 10^{-3.05} \text{ mol/L.}$$

To solve this on the calculator, locate the $\boxed{y^x}$ key. Enter 10 $\boxed{y^x}$ 3.05 $\boxed{+/-}$. The display reads 8.912509 -04. Since the logarithm has 2 digits to the right of the decimal point, $[H_3O^+]$ has 2 significant figures.

$$[H_3O^+] = 8.9 \times 10^{-4} \text{ mol/L} \quad .$$

An alternate method for finding $[H_3O^+]$ from pH involves antilogarithms. Antilogarithms can be found using a calculator or a table of logarithms. (Appendix B in the text contains information for finding antilogarithms using a table.)

$$pH = -\log [H_3O^+]$$

Multiply both sides of the equation by negative one.

$$-pH = \log [H_3O^+]$$
$$-3.05 = \log [H_3O^+]$$

Take the antilogarithm of both sides of the equation.

$$\text{antilog } (-3.05) = \text{antilog } (\log [H_3O^+])$$
$$\text{antilog } (-3.05) = [H_3O^+]$$

On the calculator, locate the inverse key $\boxed{\text{INV}}$.  When $\boxed{\text{INV}}$ $\boxed{\text{log}}$ is entered, the antilog of a number can be found.

Press 3.05 $\boxed{+/-}$ $\boxed{\text{INV}}$ $\boxed{\text{log}}$.

The display reads 8.912509 -04.

$$[H_3O^+] = 8.9 \times 10^{-4} \text{ mol/L}$$

b.  pOH = 7.33

First, find pH.
$$pH + pOH = 14.00$$
$$pH = 14.00 - 7.33$$
$$= 6.67$$

Then calculate $[H_3O^+]$.
$$[H_3O^+] = 10^{-pH} \text{ mol/L}$$
$$[H_3O^+] = 10^{-6.667} \text{ mol/L}$$

Press 10 $\boxed{y^x}$ 6.67 $\boxed{+/-}$.

The display reads 2.137962 -07.

$$[H_3O^+] = 2.14 \times 10^{-7} \text{ mol/L.}$$

Alternate method:  
$$pH = -\log [H_3O^+]$$
$$-pH = \log [H_3O^+]$$
$$-6.67 = \log [H_3O^+]$$
$$\text{antilog } (-6.67) = [H_3O^+]$$

Press 6.67 $\boxed{+/-}$ $\boxed{\text{INV}}$ $\boxed{\text{log}}$.

$$[H_3O^+] = 2.14 \times 10^{-7} \text{ mol/L}$$

Consult Appendix C in the text for answers to parts c and d.

***Practice Problem B

What is the $[H_3O^+]$ of the following solutions?

a.  $[OH^-] = 7.5 \times 10^{-6}$ M

b.  $1.5 \times 10^{-2}$ M KOH

c.   pH = 8.95

d.   pOH = 2.40

26.   A 0.35 M solution of a weak acid, HA, has a pH of 3.47. What is $K_a$ for the acid?

First, write the equation for the ionization of the weak acid in water:

$$HA_{(aq)} + H_2O_{(l)} \rightleftharpoons H_3O^+_{(aq)} + A^-_{(aq)}$$

Then to find $[H_3O^+]$ at equilibrium, convert pH to $[H_3O^+]$.

$$[H_3O^+] = 10^{-pH} \text{ mol/L}$$
$$= 10^{-3.47} \text{ mol/L}$$
$$= 3.4 \times 10^{-4} \text{ mol/L}$$

Next, calculate [HA] at equilibrium. For weak acids, $[H_3O^+]$ is the same value numerically as the molarity of HA that dissociated,

$$[H_3O^+] = [HA]_{dissociated}$$

because each mole of HA that dissociates produces one mole of $H_3O^+$ (and one mole of $A^-$).

$$[HA]_{equilibrium} = [HA]_{initial} - [HA]_{dissociated}$$
$$= 0.35 \text{ M} - 3.4 \times 10^{-4} \text{ M}$$
$$= 0.35 \text{ M} - 0.00034 \text{ M}$$

The concentration of [HA] dissociated is negligible, since HA is a weak acid.

$$[HA]_{equilibrium} = 0.35 \text{ M}$$

Finally, calculate $K_a$.

The equilibrium concentration of $[H_3O^+]$ is equal to $[A^-]$, because 1 mole of HA that dissociates produces 1 mole each of $H_3O^+$ and $A^-$.

$$K_a = \frac{[H_3O^+][A^-]}{[HA]}$$

$$K_a = \frac{(3.4 \times 10^{-4})(3.4 \times 10^{-4})}{0.35}$$
$$= 3.3 \times 10^{-7}$$

*****Practice Problem C**

    a.  0.10 M solution of hypochlorous acid, HClO, has a pH of 4.23. What is the $K_a$ for this acid?

28.    Lactic Acid, $HC_3H_5O_3$, forms in muscles during vigorous exercise. If a 0.25 M aqueous solution of lactic acid ionizes to the extent of 2.4%, calculate the pH of the solution, and find the $K_a$ for lactic acid.

First, write the equation for the ionization of lactic acid in water:

$$HC_3H_5O_3 + H_2O \rightleftharpoons H_3O^+ + C_3H_5O_3^-$$

Then, find $[H_3O^+]$.

    If 2.4% of $HC_3H_5O_3$ ionizes,

$$[HC_3H_5O_3]_{dissociated} = (0.024)[HC_3H_5O_3]_{initial}$$
$$= (0.024)(0.25 \text{ M})$$
$$= 0.0060 \text{ M}$$
$$[HC_3H_5O_3]_{dissociated} = [H_3O^+] = 6.0 \times 10^{-3} \text{ m}$$

There is now enough information to calculate pH and $K_a$. This can be done in any order.

To calculate pH,

$$\begin{aligned} pH &= -\log [H_3O^+] \\ &= -\log (6.0 \times 10^{-3}) \\ &= -(\log 6.0 + \log 10^{-3}) \\ &= -(0.78 - 3) \\ &= 2.22 \end{aligned}$$

To calculate $K_a$,

$$\begin{aligned} K_a &= \frac{[H_3O^+][C_3H_5O_3^-]}{[HC_3H_5O_3]} \\ &= \frac{(6.0 \times 10^{-3})(6.0 \times 10^{-3})}{(0.25)} \\ &= 1.4 \times 10^{-4} \end{aligned}$$

***Practice Problem D

A 0.10 M aqueous solution of acetic acid, $HC_2H_3O_2$, ionizes to the extent of 1.3%.  Calculate $[H_3O^+]$, pH and $K_a$ for this acetic acid solution.

30.    Predict whether an aqueous solution of each of the following will be acidic, basic, or neutral.  If you think the solution will be acidic or basic, write an equation for the reaction that occurs.

a.   $KNO_2$

The cation of the salt comes from a base, and the anion comes from an acid.  Separate the salt into cation and anion and determine the formulas of the original acid and base.  In most cases the base formula is determined by combining $OH^-$ with the cation.  In all cases the acid formula is determined by combining $H^+$ with the anion.

The base containing $K^+$ is KOH.  The acid containing $NO_2^-$ is $HNO_2$.

Identify the original acid and base as strong or weak (see Chapter 14).  KOH is a strong base.  $HNO_2$ is a weak acid.

Determine which ions (if any) hydrolyze (see section 15.7 in the summary outline).  Ions of weak acids or bases will hydrolyze in water.  This happens because the ion reacts with water to form the original acid or base.  That acid or base is weak, so equilibrium lies far to the right.  The weak species formed will not ionize to any extent.  For $NO_2^-$,

$$NO_2^-_{(aq)} + H_2O_{(l)} \rightleftharpoons HNO_2{}_{(aq)} + OH^-_{(aq)}$$

This reaction has little tendency to go the left.  The hydrolysis of $NO_2^-$ results in the formation of $OH^-$.  The solution is basic.

In contrast, $K^+$ does not hydrolyze.

$$K^+_{(aq)} + 2 H_2O_{(l)} \xrightarrow{\quad X \quad} KOH + H_3O^+_{(aq)}$$

Since KOH is a strong base, there is no tendency for the reaction to the right to occur.  KOH would ionize completely.  No $H_3O^+$ forms, and the pH of the solution is not affected by the presence of $K^+$ ions.

b.  $NH_4Br$

First, separate the salt into ions and determine the formulas of the original acid and base.

$$NH_4^+ \quad , \text{ original base } NH_3$$
$$Br^- \quad , \text{ original acid } HBr$$

Then, identify the acid and base as strong or weak.  Predict from this which ions will hydrolyze.  $NH_3$ is a weak base, so $NH_4^+$ hydrolyzes.  HBr is a strong acid, so $Br^-$ will not hydrolyze.

Finally, write the reaction for the ion that hydrolyzes.  Examine the product to determine if the solution is acidic or basic.

$$NH_4^+ + H_2O \rightleftharpoons NH_3 + H_3O^+$$

The aqueous solution of $NH_4^+$ is acidic since $H_3O^+$ is formed.

Consult Appendix C in the text for answers to parts c and d.

***Practice Problem E

$NH_4NO_3$ is a fertilizer.  How is the pH of the soil changed by the application of an aqueous solution of $NH_4NO_3$?

---

SELF-TEST

---

Circle the correct answer in the following multiple choice questions.

1.  Which of the following are classified as weak electrolytes?

    a.  insoluble ionic compounds
    b.  molecules
    c.  $HC_2H_3O_2$, $HNO_2$, HF
    d.  HCl, $H_2SO_4$, $HNO_3$

2. Which of the following is characteristic of a basic solution?

   a. pH = 6
   b. pOH = 10
   c. $[H_3O^+] = 1 \times 10^{-11}$ M
   d. $[OH^-] = 1 \times 10^{-11}$ M

3. The pH of a solution containing $[H_3O^+] = 2.5 \times 10^{-3}$ M, is

   a. 11.40
   b. 2.60
   c. -11.40
   d. -2.60

4. A $2.0 \times 10^{-3}$ M solution of HCl has a pH of

   a. 2.70
   b. 3.00
   c. 11.0
   d. 11.30

5. The pH of a 0.0050 M solution of $Ca(OH)_2$ is

   a. 2.00
   b. 2.30
   c. 12.00
   d. 11.70

6. What is the $H_3O^+$ in a solution whose pH is 8.50?

   a. $9.3 \times 10^{-1}$ M
   b. $3.2 \times 10^{-9}$ M
   c. $5.0 \times 10^{-9}$ M
   d. 5.5 M

7. A 1.0 M solution of HCNO, a weak acid, has a pH of 1.82. The $K_a$ for HCNO is

   a. $2.3 \times 10^{-4}$
   b. $1.5 \times 10^{-2}$
   c. $4.4 \times 10^{-25}$
   d. $6.6 \times 10^{-13}$

8. A 1.0 M solution of $HCHO_2$ ionizes to the extent of 1.3%  The pH of the solution is

   a. 1.30
   b. 2.00
   c. 12.11
   d. 1.89

9. Which of the following salts hydrolyzes to form a basic solution?

   a. $CaCl_2$
   b. $Ca_3(PO_4)_2$
   c. $NH_4Cl$
   d. $(NH_4)_3PO_4$

10.   Which of the following pairs of substances form buffer
      solutions when equimolar quantities are dissolved in water?

      a.   $CaSO_4$ and $H_2SO_4$
      b.   $KNO_3$ and $HNO_3$
      c.   NaCN and HCN
      d.   $NH_4Cl$ and HCl

---

## ANSWERS TO PRACTICE PROBLEMS

Practice Problem A

a.
$$\log [H_3O^+] = \log (2.6 \times 10^{-3})$$
$$= \log 2.6 + \log 10^{-3}$$
$$= 0.41 + (-3)$$
$$= -2.59$$
$$pH = -\log [H_3O^+]$$
$$= -(-2.59) = 2.59$$

b.
$$\log [OH^-] = \log (2.6 \times 10^{-3})$$
$$= \log 2.6 + \log 10^{-3}$$
$$= 0.41 + (-3)$$
$$= -2.59$$
$$pOH = 2.59$$

$$pH + pOH = 14$$
$$pH = 14 - pOH = 14 - 2.59$$
$$= 11.41$$

c.   $HNO_3$ is a strong, monoprotic acid.  $HNO_3$ ionizes com-
     pletely in water according to the following reaction.

$$HNO_{3(aq)} + H_2O_{(1)} \longrightarrow H_3O^+_{(aq)} + NO_3^-_{(aq)}$$

$[H_3O^+]$ is the same value numerically as the molarity of
$HNO_3$.

$$[H_3O^+] = 5.5 \times 10^{-2} \text{ M}$$
$$\log [H_3O^+] = \log (5.5 \times 10^{-2}$$
$$= \log 5.5 + \log 10^{-2}$$
$$= 0.74 + (-2)$$
$$= -1.26$$
$$pH = -\log [H_3O^+]$$
$$= -(-1.26) = 1.26$$

d.   In solution with water, $Ba(OH)_2$ is a strong base.
     $Ba(OH)_2$ ionizes completely according to the following
     reaction:

$$Ba(OH)_{2(s)} \xrightarrow{H_2O} Ba^{2+}_{(aq)} + 2 \, OH^-_{(aq)}$$

A $5.5 \times 10^{-2}$ M solution of $Ba(OH)_2$ produces a
$2(5.5 \times 10^{-2}$ M) concentration of $OH^-$.

$$[OH^-] = 1.1 \times 10^{-1}\ M$$
$$\log\,[OH^-] = \log\,(1.1 \times 10^{-1})$$
$$= \log 1.1 + \log 10^{-1}$$
$$= 0.041 + (-1)$$
$$= -0.959$$
$$pOH = -\log\,[OH^-]$$
$$= -(-0.959) = 0.959$$
$$pH + pOH = 14$$
$$pH = 14 - pOH$$
$$= 14 - 0.959$$
$$= 13.04$$

**Practice Problem B**

a.
$$[H_3O^+][OH^-] = K_w$$
$$[H_3O^+] = \frac{K_w}{[OH^-]} = \frac{1.00 \times 10^{-14}}{7.5 \times 10^{-6}}$$
$$= 1.3 \times 10^{-9}\ M$$

b.  KOH is a strong base.  In water, $1.5 \times 10^{-2}$ M KOH ionizes to form $1.5 \times 10^{-2}$ M $OH^-$.

$$[H_3O^+][OH^-] = K_w$$
$$[H_3O^+] = \frac{K_w}{[OH^-]} = \frac{1.00 \times 10^{-14}}{1.5 \times 10^{-2}}$$
$$= 6.7 \times 10^{-13}$$

c.
$$[H_3O^+] = 10^{-pH}\ mol/L.$$
$$[H_3O^+] = 10^{-8.98}$$
$$= 1.1 \times 10^{-9}\ mol/L$$

Alternate method:
$$pH = -\log\,[H_3O^+]$$
$$-pH = \log\,[H_3O^+]$$
$$-8.95 = \log\,[H_3O^+]$$

$$antilog\,(-8.95) = [H_3O^+]$$
$$1.1 \times 10^{-9}\ mol/L = [H_3O^+]$$

d.
$$pH + pOH = 14$$
$$pH = 14 - pOH$$
$$pH = 14 - 2.40$$
$$= 11.60$$
$$[H_3O^+] = 10^{-pH}\ mol/L$$
$$= 10^{-11.60}$$
$$= 2.5 \times 10^{-12}\ mol/L$$

Alternate method:
$$pH = -\log\,[H_3O^+]$$
$$-pH = \log\,[H_3O^+]$$
$$-11.60 = \log\,[H_3O^+]$$

$$antilog\,(-11.60) = [H_3O^+]$$
$$2.5 \times 10^{-12}\ mol/L = [H_3O^+]$$

**Practice Problem C**

The ionization of HClO is shown in the following equation:

$$HClO_{(aq)} + H_2O_{(l)} \rightleftharpoons H_3O^+_{(aq)} + ClO^-_{(aq)}$$

The $[H_3O^+]$ at equilibrium is calculated from pH.

$$[H_3O^+] = 10^{-pH} \text{ mol/L}$$
$$= 10^{-4.23}$$
$$= 5.9 \times 10^{-5} \text{ mol/L}$$

$$[HClO] \text{ equilibrium} = 0.10 \text{ M} - 5.9 \times 10^{-5} \text{ M}$$
$$= 0.10 \text{ M}$$

$$[ClO^-] \text{ equilibrium} = [H_3O^+] \text{ equilibrium}$$
$$= 5.9 \times 10^{-5} \text{ M}$$

$$K_a = \frac{[H_3O^+][Cl^-]}{[HClO]}$$
$$= \frac{(5.9 \times 10^{-5})(5.9 \times 10^{-5})}{(0.10)}$$
$$= 3.5 \times 10^{-8}$$

**Practice Problem D**

First write the equation for the ionization of acetic acid.

$$HC_2H_3O_2 + H_2O \rightleftharpoons H_3O^+ + C_2H_3O_2^-$$

$$[H_3O^+] = [HC_2H_3O_2]_{dissociated} = (0.013)[HC_2H_3O_2]_{initial}$$
$$= (0.013)(0.10)$$
$$= 0.0013 \text{ M} = 1.3 \times 10^{-3} \text{ M}$$
$$pH = -\log [H_3O^+]$$
$$= -\log (1.3 \times 10^{-3})$$
$$= -(-2.89)$$
$$= 2.89$$

$$K_a = \frac{[H_3O^+][C_2H_3O_2^-]}{[HC_2H_3O_2]}$$
$$= \frac{(1.3 \times 10^{-3}(1.3 \times 10^{-3})}{(0.10)}$$
$$= 1.7 \times 10^{-5}$$

**Practice Problem E**

$NH_4^+$ is the cation of the weak base, $NH_3$. $NO_3^-$ is the anion of the strong acid, $HNO_3$. $NH_4^+$ will hydrolyze

$$NH_4^+{}_{(aq)} + H_2O_{(l)} \rightleftharpoons NH_3{}_{(aq)} + H_3O^+_{(aq)}$$

The formation of $H_3O^+$, an acid, lowers the pH of the soil, making it more acidic.

---

ANSWERS TO SELF-TEST

1.  c
2.  c
3.  b
4.  a
5.  c
6.  b
7.  a
8.  d
9.  b
10. c

# OXIDATION AND REDUCTION

**CHAPTER 16**

---

SUMMARY OUTLINE

---

16.1    Concepts of Oxidation and Reduction

An oxidation-reduction reaction is a reaction that involves changes in oxidation states; one substance is oxidized, another is reduced.

Oxidation is defined as a loss of electrons and therefore an increase in oxidation number.

Reduction is defined as a gain of electrons and therefore a decrease in oxidation number.

An oxidizing agent, or oxidant, causes oxidation by accepting electrons from the other reactant.  The oxidizing agent is reduced.

The reducing agent, or reductant, causes reduction by supplying electrons for the other reactant to accept.  The reducing agent is oxidized.

Ions need not be present in a redox reaction.  the only requirement is that there is an increase in oxidation number at the same time that there is a decrease in oxidation number.

16.2    Balancing Redox Equations:  The Oxidation State Method

The oxidation state method is used for nonionic equations to balance the total increase in oxidation state with the total decrease in oxidation state.

Five steps are used in the oxidation state method:

1. Write correct formulas for reactants and products, and assign oxidation numbers to each element. Identify the element(s) that undergo changes in oxidation numbers and balance their atoms.
2. Draw a bridge to connect reactant and product for the oxidized element and for the reduced element. For each element, write the <u>total</u> change in oxidation state on the bridge.
3. Balance the total increase in oxidation state with the total decrease in oxidation state, multiplying the changes in oxidation states by small whole number coefficients to achieve the least common multiple.
4. Use the multipliers from step 3 as coefficients of the respective elements or compounds in the equation.
5. Balance remaining elements by inspection.

## 16.3  <u>Balancing Redox Equations:  The Ion-Electron Method</u>

The ion-electron method is used for net ionic equations, and it balances the numbers of electrons transferred in the oxidation and reduction half reactions.

The steps used in the ion-electron method depend on whether the reaction takes place in acidic or basic solution. For acidic solutions:

1. Assign oxidation numbers to each element. Write the oxidation half-reaction and the reduction half-reaction and balance the elements undergoing oxidation and reduction.
2. Balance oxygen in the half-reactions by adding $H_2O$ to the side of the equation that is deficient in oxygen. One $H_2O$ is needed for each atom.
3. Balance hydrogen by adding $H^+$ to the side of the equation that is deficient in hydrogen.
4. Add electrons to each half reaction to balance the electrical charges. (The $e^-$ will appear on the right in an oxidation reaction and on the left in a reduction reaction.)
5. Balance the electrons lost with the electrons gained. Multiply the half-reactions by the smallest whole number that balances the electron transfer.
6. Add the two half-reactions together and cancel any identical species found on both sides of the equation.

For reactions in basic solution, steps 2 and 3 are changed to show the basic species in solution. The changes for balancing ionic equations for basic solutions by the ion-electron method are:

2. Balance oxygen in the half-reactions by adding $OH^-$ ions to the side of the equation that is deficient in oxygen. <u>Two</u> $OH^-$ ions are needed for each O atom.
3. Balance hydrogen by adding $H_2O$ to the side of the equation that is deficient in hydrogen.

16.4   <u>The Activity Series of Metal</u>

The activity series of metals list metals in decreasing order of ability to displace hydrogen from acids and thus become oxidized (Table 16-1).

$$2M + 2\ H^+ \longrightarrow 2\ M^+ + H_2$$

$$(2\ M \longrightarrow 2\ M^+ + 2\ e^-;\ oxidation)$$

Atoms of a metal will spontaneously lose electrons to cations of any other metal (or hydrogen) listed below it in the activity series.  These reactions, illustrated below, are called metal displacement reaction.

$$nA + mB^{n+} \longrightarrow nA^{m+}\ mB$$

where metal A is higher than metal B in the activity series.

16.5   <u>Voltaic Cells</u>

Electrochemical reactions are reactions capable of producing electrical energy.  The apparatus in which an electrochemical reaction is run is called an electrochemical cell.  The electrochemical cell produces useable energy if the oxidation half-reaction takes place in a separate container from the reduction half-reaction.  Each half-cell contains a solution of a strong electrolyte and an electrically conductive electrode.  The two electrodes are connected by a metal wire.  A salt bridge, connecting the two solutions, maintains electrical neutrality in the solutions (Figure 16-3).

In a voltaic cell, the electrode where oxidation occurs is called the anode, the electrode where reduction occurs is called the cathode.  The anode conducts electrons away from the solution into the wire and the cathode takes electrons from the wire and provides them to the solution.

Electrochemical cells that provide electricity are called voltaic cells or galvanic cells.

16.6   <u>Batteries</u>

Batteries are commercial voltaic cells used as sources of electricity.  The most common types are the lead storage battery (Figure 16-4), the dry cell (Figure 16-5) and the nickel-cadmium battery.

16.7   <u>Electrolytic Cells</u>

When electrical energy from an outside source is applied to a nonspontaneous electrochemical reaction, a reaction will occur.  This process, called electrolysis, occurs in an electrolytic cell (Figure 16-6).

In the electrolytic cell, electrons flow from the outside source to the cathode.  The electric current is carried from the cathode to the anode by ions.  The anode sends electrons back to the outside electrical source.

The cathode is negatively charged because it receives electrons.  Cations are attracted to the cathode where they accept electrons, undergoing reduction.

The anode attracts anions.  The anions lose electrons to the anode as they undergo oxidation.  The electrons are transmitted back to the electrical source.

---

### SOLUTIONS TO SELECTED STUDY QUESTIONS AND PROBLEMS
### and
### PRACTICE PROBLEMS

---

3.  For each reaction, identify the substance that is oxidized, the substance that is reduced, the oxidizing agent, and the reducing agent.

   a.  $3 P + 5 HNO_3 + 2 H_2O \longrightarrow 5 NO + 3 H_3PO_4$

   First, assign oxidation numbers to each element according to the guidelines given on page 89 in the study guide.  Write the oxidation number below the symbol, using a + or − charge before the number.

$$3 P + 5 HNO_3 + 2 H_2O \longrightarrow 5 NO + 3 H_3PO_4$$
$$0 \qquad +1+5-2 \qquad +1-2 \qquad +2-2 \qquad +1+5-2$$

   Next identify elements whose oxidation states have changed.

   P has a change in oxidation state from 0 to +5.  N has a change in oxidation state from +5 to +2.

   Then, determine which change represents oxidation and which represents reduction.  A useful aid for determining which element is oxidized (reducing agent) and which element is reduced (oxidizing agent) is a simple number line.  Label the direction for oxidation (increase in oxidation number and the direction for reduction (decrease in oxidation number.)  Then write the elements on the line according to their oxidation numbers as reactants and products.  Draw an arrow from the reactant oxidation state to the product oxidation state.

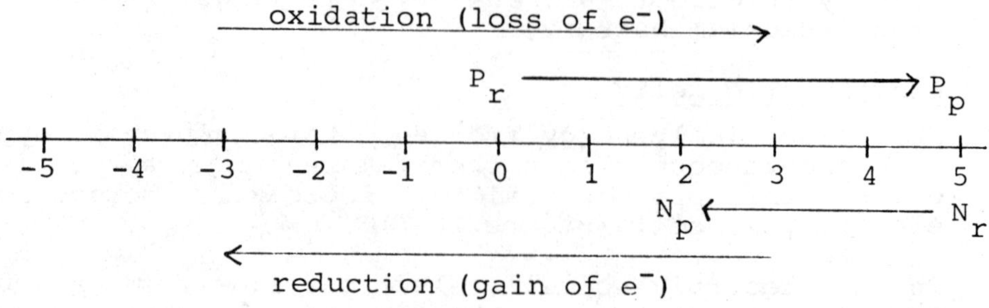

The direction of the arrow graphically shows that P is oxidized (reducing agent) and N is reduced (oxidizing agent).

b.   $Sn + 4\ HNO_3 \longrightarrow SnO_2 + 4\ NO_2 + H_2O$

  0        +1+5-2        +4-2        +4-2      +1-2

Sn has a change in oxidation state from 0 to +4.  N has a change in oxidation state from +5 to +4.

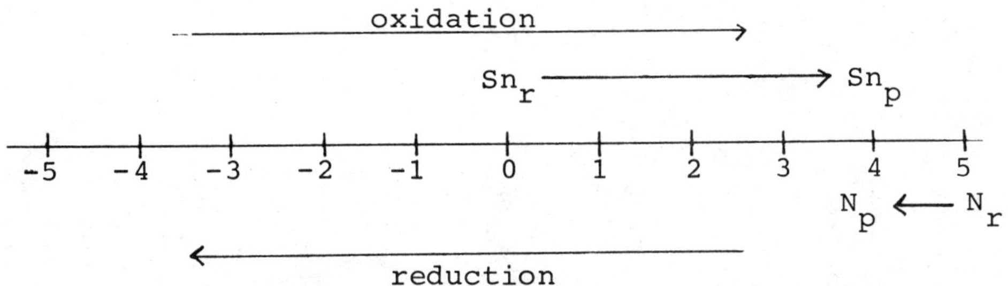


Sn is oxidized (reducing agent) and N is reduced (oxidizing agent).

c.   $I_2 + 5\ Cl_2 + 6\ H_2O \longrightarrow 2\ HIO_3 + 10\ HCl$

  0         0          +1 -2          +1+5-2        +1-1

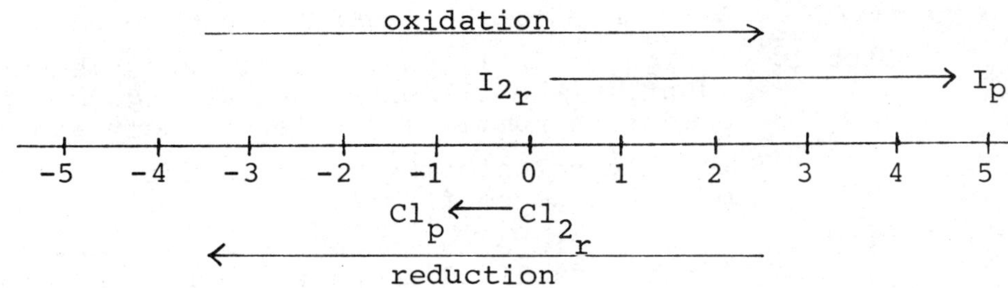


$I_2$ is oxidized (reducing agent) and $Cl_2$ is reduced (oxidizing agent).

Part d is worked in the same way as parts a, b, and c.

***Practice Problem A

Identify the substance that is oxidized, the substance that is reduced, the oxidizing agent and the reducing agent in the following reactions.

a.   $2\ KClO_3 \longrightarrow 2\ KCl + 3\ O_2$

b.   $NH_3 + 3\ O_2 \longrightarrow 2\ N_2 + 6\ H_2O$

4.      Use the oxidation state method to balance the following equations.

a.    $KMnO_4 + -2\ KCl + H_2SO_4 \longrightarrow MnSO_4 + K_2SO_4 + H_2O + Cl_2$
     +1+7−2        +1−1   +1+6−2        +2+6−2     +1+6−2     +1−2    0

Step 1.   Assign oxidation numbers to each element. Identify the elements that undergo changes in oxidation number and balance their atoms.

To balance chlorine use a coefficient 2 in front of KCl.  No coefficient is needed for Mn.

Step 2.   Draw a bridge to connect reactant and product for the oxidized element and for the reduced element.  Write the _total_ change in oxidation number.

$$
\begin{array}{l}
\quad\quad\quad\quad\quad\quad\quad\quad +2 \\
\quad\quad\quad\quad -5 \\
KMnO_4 + 2\ KCl + H_2SO_4 \longrightarrow MnSO_4 + K_2SO_4 + H_2O + Cl_2 \\
\ \ +7 \quad\quad\ -1 \quad\quad\quad\quad\quad\quad +2 \quad\quad\quad\quad\quad\quad\quad\quad\quad 0
\end{array}
$$

Cl has a +2 total change in oxidation state since 2 chlorine atoms each have a +1 change.

Step 3.   Balance the total increase in oxidation state with the total decrease in oxidation state. The least common multiple of +2 and -5 is 10.

$$\underset{\text{5(+2)}}{\underset{\text{2(-5)}}{KMnO_4 + 2KCl + H_2SO_4 \longrightarrow MnSO_4 + K_2SO_4 + H_2O + Cl_2}}$$

Step 4.   Use the multipliers in step 3 as coefficients in the equation. In the case of KCl, 5 must be multiplied by the coefficient already present.

$$2\ KMNO_4 + 10\ KCl + H_2SO_4 \longrightarrow 2\ MnSO_4 + K_2SO_4 + H_2O + 5\ Cl_2$$

Step 5.   Balance remaining elements by inspection. There are 12 K's on the left (2 in 2 $KMnO_4$ plus 10 in 10 KCl); place a 6 in front of $K_2SO_4$. Count $SO_4$'s on right. There are 8 $SO_4$'s (2 in 2 $MnSO_4$ plus 6 in 6 $K_2SO_4$); place an 8 in front of $H_2SO_4$ on the left. This gives 16 H's on left; place an 8 in front of $H_2O$ on the right. Count O's not found in $SO_4$. On the left, there are 8 in 2 $KMnO_4$ and on the right there are 8 in 8 $H_2O$. The equation is balanced.

$$2\ KMnO_4 + 10\ KCl + 8\ H_2SO_4 \longrightarrow 2\ MnSO_4 + 6\ K_2SO_4 + 8\ H_2O + 5\ Cl_2$$

b.   $H_2O + P_4 + HClO \longrightarrow H_3PO_4 + HCl$

Once the principles of balancing equations by the oxidation state method are understood, steps 1-4 can be combined. The equation below illustrates this.

$$\underset{\text{+1-2}\quad\text{0}\quad\text{+1+1}\quad\text{+1+5-2}\quad\text{+1-1}}{\underset{\text{+20}}{\overset{\boxed{10}\;\boxed{-2}}{H_2O + P_4 + 10\ HClO \longrightarrow 4\ H_3PO_4 + 10\ HCl}}}$$

The arrows connected to $\boxed{10}$ show the location for the placement of the coefficients. The remaining elements are balanced by inspection.

$$6\ H_2O + P_4 + 10\ HClO \longrightarrow 4\ H_3PO_4 + 10\ HCl$$

c.   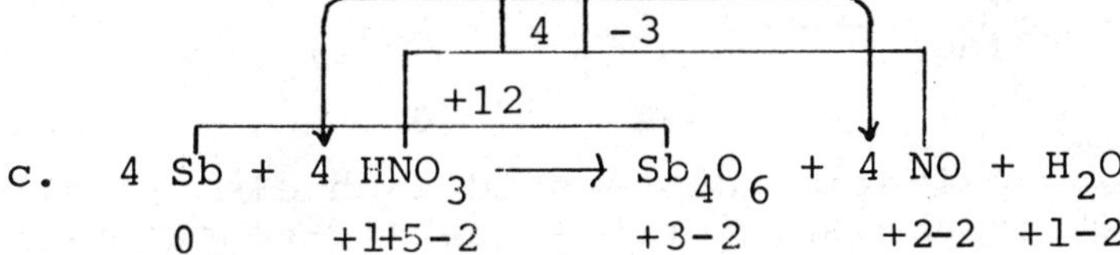

$$\underset{\quad 0 \qquad\ +1+5-2 \qquad +3-2 \qquad\ +2-2\ \ +1-2}{\underset{\text{+12}}{\overset{\boxed{4}\;\boxed{-3}}{4\ Sb + 4\ HNO_3 \longrightarrow Sb_4O_6 + 4\ NO + H_2O}}}$$

By inspection,

$$4\ Sb + 4\ HNO_3 \longrightarrow Sb_4O_6 + 4\ NO + 2\ H_2O$$

d. $PbO_2 + 2\ HI \longrightarrow PbI_2 + I_2 + H_2O$

+4-2    +1-1    +2 -1    0    +1-2

By inspection,

$$PbO_2 + 4\ HI \longrightarrow PbI_2 + I_2 + 2\ H_2O$$

***Practice Problem B

Balancing the following equations by the oxidation state method:

a. $Na_2TeO_3 + NaI + HCl \longrightarrow NaCl + Te + H_2O + I_2$

b. $H_2S + NHO_3 \longrightarrow S + NO + H_2O$

7.    Use the ion-electron method to balance the following equations. All reactions occur in acidic solution.

a. $AsH_{3(g)} + Ag^+_{(aq)} \longrightarrow As_4O_{6(s)} + Ag_{(s)}$

    -3+1       +1       +3 -2      0

Step 1:  Assign oxidation numbers to each element (see equation above). Write the half-reactions, balancing the elements undergoing oxidation and reduction.

$$4 \text{ AsH}_3 \xrightarrow{\phantom{xxx}} \text{As}_4\text{O}_6 \quad \text{(oxidation)}$$
$$\phantom{4 \text{ A}}{-3} \phantom{xxxxxxx} {+3}$$

(As has an increase in oxidation number.)

$$\text{Ag}^+ \longrightarrow \text{Ag} \quad \text{(reduction)}$$
$$+1 \phantom{xxxxx} 0$$

(Ag has a decrease in oxidation number.)

Step 2:  Balance O in each half-reaction by adding $H_2O$.

$$6 \text{ H}_2\text{O} + 4 \text{ AsH}_3 \longrightarrow \text{As}_4\text{ O}_6 \quad \text{(oxidation}$$

$$\text{Ag}^+ \longrightarrow \text{Ag} \quad \text{(reduction)}$$

Step 3:  Balance H in each half-reaction by adding $H^+$.

$$6 \text{ H}_2\text{O} + 4 \text{ AsH}_3 \longrightarrow \text{As}_4\text{O}_6 + 24 \text{ H}^+ \quad \text{(oxidation)}$$

$$\text{Ag}^+ \longrightarrow \text{Ag} \quad \text{(reduction)}$$

Step 4:  Add electrons to each half-reaction to balance the electrical charge.

$$6 \text{ H}_2\text{O} + 4 \text{ AsH}_3 \longrightarrow \text{As}_4\text{O}_6 + 24 \text{ H}^+ + 24 \text{ e}^- \quad \text{(oxidation)}$$

(charge)  $0 \phantom{xxxx} 0 \phantom{xxxxxxxx} 0 \phantom{xxx} +24 \phantom{xxx} -24$

$$0 = 0$$

$$1 \text{ e}^- + \text{Ag}^+ \longrightarrow \text{Ag} \quad \text{(reduction)}$$

(charge)  $-1 \phantom{xxx} +1 \phantom{xxxxx} 0$

$$0 = 0$$

Step 5:  Balance electrons lost with electrons gained.

$$6 \text{ H}_2\text{O} + 4 \text{ AsH}_3 \longrightarrow \text{As}_4\text{O}_6 + 24 \text{ H}^+ + 24 \text{ e}^- \quad \text{(oxidation)}$$

$$24[1 \text{ e}^- + \text{Ag}^+ \longrightarrow \text{Ag}] \quad \text{(reduction)}$$

Step 6:  Add the two half-reactions.

$$6 \text{ H}_2\text{O} + 4 \text{ AsH}_3 + 24 \cancel{\text{e}^-} + 24 \text{ Ag}^+ \longrightarrow \text{As}_4\text{O}_6 + 24 \text{ H}^+ + 24 \cancel{\text{e}^-} + 24 \text{ Ag}$$

$$6\text{H}_2\text{O}_{(1)} + 4 \text{ AsH}_{3(g)} + 24 \text{ Ag}^+_{(aq)} \longrightarrow \text{As}_4\text{O}_{6(s)} + 24 \text{ H}^+_{(aq)} + 24 \text{ Ag}_{(s)}$$

b.  $\text{Zn}_{(s)} + \text{H}_2\text{MoO}_{4(aq)} \longrightarrow \text{Zn}^{2+}_{(aq)} + \text{Mo}^{3+}_{(aq)}$

$\phantom{b.}\; 0 \phantom{xxxx} +1+6-2 \phantom{xxxxxx} +2 \phantom{xxxxx} +3$

Step 1:  (See oxidation numbers above)

$$\text{Zn} \longrightarrow \text{Zn}^{2+} \quad \text{(oxidation)}$$
$$0 \phantom{xxxxx} +2$$

$$\text{H}_2\text{MoO}_4 \longrightarrow \text{Mo}^{3+} \quad \text{(reduction)}$$
$$+6 \phantom{xxxxx} +3$$

Step 2:  $\quad \text{Zn} \longrightarrow \text{Zn}^{2+} \quad \text{(oxidation)}$

$$\text{H}_2\text{MoO}_4 \longrightarrow \text{Mo}^{3+} + 4 \text{ H}_2\text{O} \quad \text{(reduction)}$$

Step 3: $\quad$ $Zn \longrightarrow Zn^{2+}$ $\hspace{2cm}$ (oxidation)

$\quad 6\ H^+ + H_2MoO_4 \longrightarrow Mo^{3+} + 4\ H_2O$ $\hspace{1cm}$ (reduction)

Step 4: $\quad$ $Zn \longrightarrow Zn^{2+} + 2\ e^-$ $\hspace{1.5cm}$ (oxidation)
$$0 \hspace{2cm} +2 \hspace{1cm} -2$$
$$0 = 0$$

$$3\ e^- + 6\ H^+ + H_2MoO_4 \longrightarrow Mo^{3+} + 4\ H_2O \hspace{1cm} \text{(reduction)}$$
$$-3 \hspace{0.5cm} +6 \hspace{1.5cm} 0 \hspace{1cm} +3 \hspace{1cm} 0$$
$$+3 = +3$$

Step 5: $\quad$ $3[Zn \longrightarrow Zn^{2+} + 2\ e^-]$ $\hspace{1cm}$ (oxidation)

$2[3\ e^- + 6\ H^+ + H_2MoO_4 \longrightarrow Mo^{3+} + 4\ H_2O]$ $\hspace{0.5cm}$ (reduction)

Step 6:

$$3\ Zn + 6\ \cancel{e^-} + 12\ H^+ + 2\ H_2MoO_4 \longrightarrow 3\ Zn^{2+} + 6\ \cancel{e^-} + 2\ Mo^3 + 8\ H_2O$$

$$3\ Zn_{(s)} + 12\ H^+_{(aq)} + 2\ H_2MoO_4{}_{(aq)} \longrightarrow 3\ Zn^{2+}_{(aq)} + 2\ Mo^{3+}_{(aq)} + 8\ H_2O_{(1)}$$

Consult Appendix C in the text for answers to parts c and d.

9. $\quad$ Use the ion-electron method to balance the following equations. All reactions occur in basic solution.

a. $\quad S^{2-}_{(aq)} + I_{2(s)} \longrightarrow SO_4^{2-}{}_{(aq)} + I^-_{(aq)}$
$$-2 \hspace{1.5cm} 0 \hspace{1.5cm} +6\ -2 \hspace{1.5cm} -1$$

Step 1: $\quad$ Same as with acidic equation. (See oxidation numbers above.)

$$S^{2-} \longrightarrow SO_4^{2-} \hspace{2cm} \text{(oxidation)}$$
$$-2 \hspace{1.5cm} +6$$

$$I_2 \longrightarrow 2\ I^- \hspace{2cm} \text{(reduction)}$$
$$0 \hspace{1.5cm} -1$$

Step 2: $\quad$ Balance O in each half-reaction by adding $\underline{2}$ OH$^-$ ions for each O. (Notice that O won't balance until Step 3 is complete.)

$$8\ OH^- + S^{2-} \longrightarrow SO_4^{2-} \hspace{1.5cm} \text{(oxidation)}$$
$$I_2 \longrightarrow 2\ I^- \hspace{2.5cm} \text{(reduction)}$$

Step 3: $\quad$ Balance H in each half-reaction by adding $H_2O$.

$$8\ OH^- + S^{2-} \longrightarrow SO_4^{2-} + 4\ H_2O \hspace{1cm} \text{(oxidation)}$$
$$I_2 \longrightarrow 2\ I^- \hspace{3cm} \text{(reduction)}$$

Step 4: $\quad$ Same as with acidic equations.

$$8\ OH^- + S^{2-} \longrightarrow SO_4^{2-} + 4\ H_2O + 8\ e^- \hspace{1cm} \text{(oxidation)}$$
$$-8 \hspace{1cm} -2 \hspace{2cm} -2 \hspace{2cm} 0 \hspace{1cm} -8$$
$$-10 = -10$$

$$2\ e^- + I_2 \longrightarrow 2\ I^- \hspace{2cm} \text{(reduction)}$$
$$-2 \hspace{1cm} 0 \hspace{2cm} -2$$

Step 5:

$$8\ OH^- + S^{2-} \longrightarrow SO_4^{2-} + 4\ H_2O + 8\ e^-$$
$$4[2\ e^- + I_2 \longrightarrow 2\ I^-]$$

Step 6:

$$8\ OH^- + S^{2-} + \cancel{8\ e^-} + 4\ I_2 \longrightarrow SO_4^{2-} = 4\ H_2O + \cancel{8\ e^-} + 8\ I^-$$

$$8\ OH^-_{(aq)} + S^{2-}_{(aq)} + 4\ I_{2(s)} \longrightarrow SO_4^{2-}{}_{(aq)} + 4\ H_2O_{(1)} + 8\ I^-_{(aq)}$$

Consult Appendix C in the text for answers to parts b, c, and d.

***Practice Problem C

Balance the following equations by the ion-electron method.

a.  $H_3PO_{2(aq)} + Cr_2O_7^{2-}{}_{(aq)} \longrightarrow H_3PO_{4(aq)} + Cr^{3+}_{(aq)}$  (acidic)

b.  $MnO_4^-{}_{(aq)} + ClO_2^-{}_{(aq)} \longrightarrow MnO_2 + ClO_4^-{}_{(aq)}$  (basic)

12.  From the activity series of metals (Table 16-1), predict which of the following reactions will occur.

a.  $Zn_{(s)} + Ni^{2+}_{(aq)} \longrightarrow Zn^{2+}_{(aq)} + Ni_{(s)}$

The reaction occurs.  Zn lies above Ni on Table 16-1 which means that Zn has a greater tendency to be oxidized (lose $e^-$) than does Ni.  The oxidation of Zn, the reaction in the forward direction,

$$Zn \longrightarrow Zn^{2+} + 2\ e^-$$

is favored over the oxidation of Ni, the reverse reaction,

$$Ni^{2+} + 2\ e^- \longleftarrow Ni$$

b. $Pb_{(s)} + Zn^{2+}_{(aq)} \longrightarrow Pb^{2+}_{(aq)} + Zn_{(s)}$

Pb lies below Zn on Table 16-1. The reaction does not occur.

c. $Fe_{(s)} + 2\ Ag^+_{(aq)} \longrightarrow Fe^{2+}_{(aq)} + 2\ Ag_{(s)}$

Fe lies above Ag on Table 16-1. The reaction occurs.

d. $Hg_{(l)} + Pb^{2+}_{(aq)} \longrightarrow Hg^{2+}_{(aq)} + Pb_{(s)}$

Hg lies below Pb on Table 16-1. The reaction does not occur.

---

## SELF-TEST

Circle the correct answer in the following multiple choice questions.

1. The element oxidized in the following reaction is

$$Fe + O_2 + HCl \longrightarrow FeCl_3 + H_2O$$

   a. iron
   b. oxygen
   c. hydrogen
   d. chlorine

2. The oxidizing agent in the following reaction is

$$CuO + NH_3 \longrightarrow Cu + N_2 + H_2O$$

   a. copper in CuO
   b. oxygen in CuO
   c. nitrogen in $NH_3$
   d. hydrogen in $NH_3$

3. Which of the following is not a redox reaction?

   a. $BaSO_4 + 4\ C \longrightarrow BaS + 4\ CO$
   b. $K_2Cr_2O_7 + 14\ HI \longrightarrow 2\ CrI_3 + 2\ KI + BI_2 + 7\ H_2O$
   c. $2\ Ag\ NO_3 + K_2SO_4 \longrightarrow Ag_2\ SO_4 + 2\ KNO_3$
   d. $H_2SO_3 + I_3^- \longrightarrow HSO_4^- + I^-$

4. The following equation has been partially balanced using the oxidation state method.

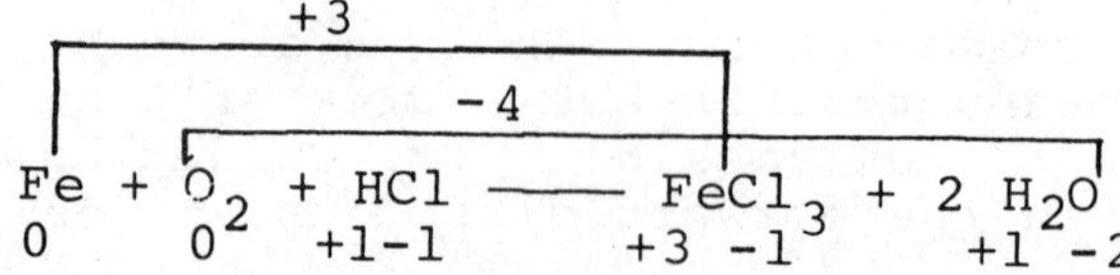

What coefficient will be found in front of Fe in the
balanced equation?

   a.  1
   b.  2
   c.  3
   d.  4

5.   How many electrons appear in the half reaction below, when
it is balanced?

$$ClO^-_{(aq)} \longrightarrow Cl^-_{(aq)} \qquad \text{(basic solution)}$$

   a.  1
   b.  2
   c.  3
   d.  4

6.   Which of the following reactions will not occur?

   a.  $K_{(s)} + Na^+_{(aq)} \longrightarrow K^+_{(aq)} + Na_{(s)}$

   b.  $3\ Cu_{(s)} + 2\ Au^{3+}_{(aq)} \longrightarrow 3\ Cu^{2+}_{(aq)} + 2\ Au_{(s)}$

   c.  $Pb_{(s)} + Zn^{2+}_{(aq)} \longrightarrow Pb^{2+}_{(aq)} + Zn_{(s)}$

   d.  $3\ Sn^{2+}_{(aq)} + 2\ Al_{(s)} \longrightarrow 3\ Sn_{(s)} + 2\ Al^{3+}_{(aq)}$

7.   Which of the following does not correctly describe a voltaic
cell?

   a.  Oxidation occurs at the anode.
   b.  A salt bridge is necessary to maintain electrical
neutrality in two separated half-cells.
   c.  Voltaic cells supply electric current.
   d.  The cathode conducts electrons away from the solution,
into the wire.

8.   In electrolytic cells,

   a.  anions are attracted to the cathode.
   b.  anions are attracted to the anode.
   c.  electricity is produced.
   d.  the anode is negatively charged.

9.   When the following reaction is balanced by the oxidation
state method,

$$PbO_2 + Sb + NaOH \longrightarrow PbO + NaSbO_2 + H_2O$$

the coefficient for $PbO_2$ is

   a.  1
   b.  2
   c.  3
   d.  4

10.   When the following acidic reaction is balanced by the ion-
electron method,

$$P_4 + NO_3^- \longrightarrow H_3PO_4 + NO$$

the number of $H_2O$ molecules appearing in the final reaction is

a.  2
b.  8
c.  16
d.  48

---

## ANSWERS TO PRACTICE PROBLEMS

**Practice Problem A**

a.  $$2\ KClO_3 \longrightarrow 2\ KCl + 3\ O_2$$
    +1+5-2         +1-1      0

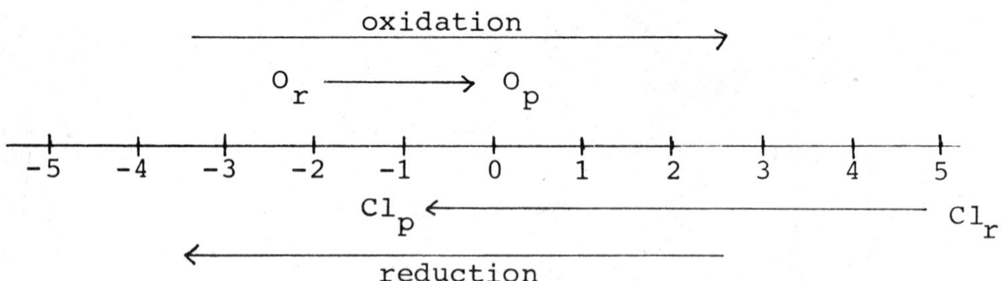


O is oxidized (reducing agent) and Cl is reduced (oxidizing agent)

b.  $$4\ NH_3 + 3\ O_2 \longrightarrow 2\ N_2 + 6\ H_2O$$
    -3+1         0              0         +1-2

Assigning oxidation numbers to ammonia, $NH_3$, may be confusing. In most formulas, the element with the lower electronegativity is written first. This first element then has a positive oxidation number. Ammonia, NH , is an exception. The guidelines on page 89 correctly predict the oxidation numbers in $NH_3$, even though the elements are reversed in the formula. (Hydrogen has an oxidation number of +1 except when combined with a less electronegative element. Nitrogen is more electronegative than hydrogen, therefore hydrogen is +1 in this compound and nitrogen is -3.)

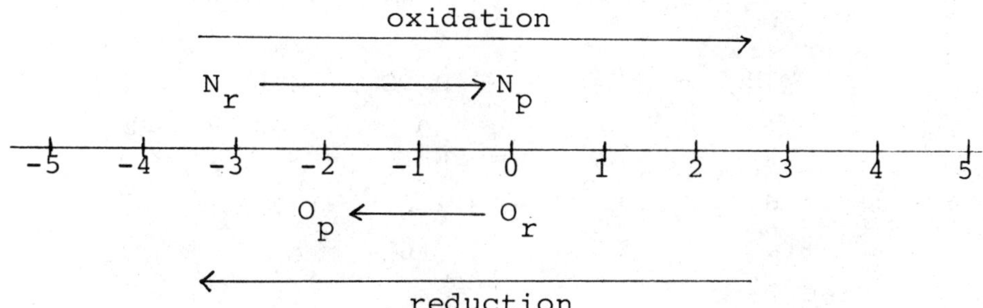


N is oxidized (reducing agent) and O is reduced
(oxidizing agent).

Practice Problem B

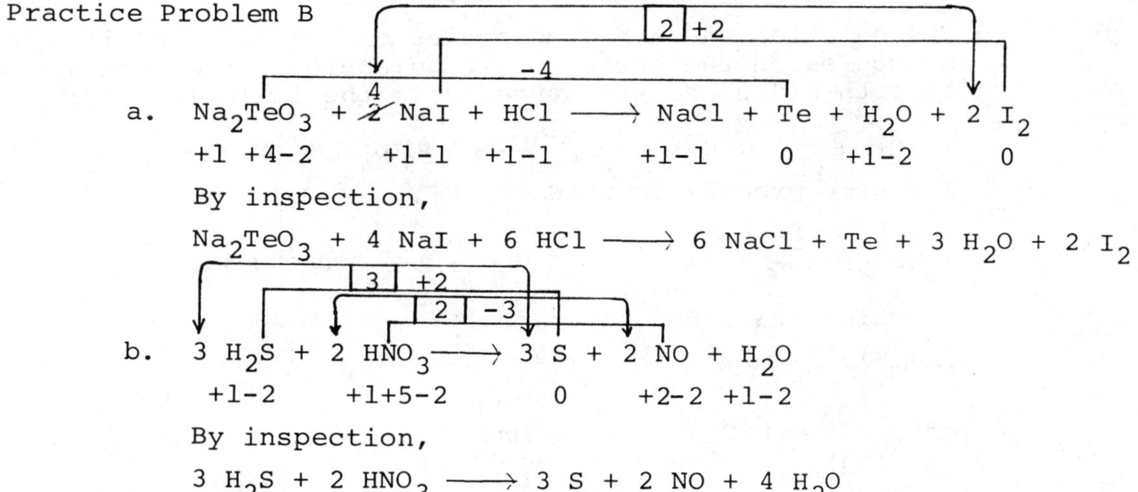

a. $Na_2TeO_3$ + 2 NaI + HCl ⟶ NaCl + Te + $H_2O$ + 2 $I_2$

   +1 +4-2    +1-1   +1-1      +1-1    0    +1-2      0

By inspection,

$Na_2TeO_3$ + 4 NaI + 6 HCl ⟶ 6 NaCl + Te + 3 $H_2O$ + 2 $I_2$

b. 3 $H_2S$ + 2 $HNO_3$ ⟶ 3 S + 2 NO + $H_2O$

   +1-2      +1+5-2         0     +2-2  +1-2

By inspection,

3 $H_2S$ + 2 $HNO_3$ ⟶ 3 S + 2 NO + 4 $H_2O$

Practice Problem C

a. $H_3PO_{2(aq)}$ + $Cr_2O_7^{2-}{}_{(aq)}$ ⟶ $H_3PO_{4(aq)}$ + $Cr^{3+}_{(aq)}$     (acidic)

   +1+1-2        +6-2            +1+5-2        +3

Step 1:   (See oxidation numbers above.)

$$H_3PO_2 \longrightarrow H_3PO_4 \qquad \text{(oxidation)}$$
$$\phantom{H_3PO_2} +1 \qquad\quad +5$$
$$Cr_2O_7^{2-} \longrightarrow 2\ Cr^{3+} \qquad \text{(reduction)}$$
$$\phantom{Cr_2O_7^{2-}} +6 \qquad\quad +3$$

Step 2:

$$2\ H_2O + H_3PO_2 \longrightarrow H_3PO_4 \qquad \text{(oxidation)}$$
$$Cr_2O_7^{2-} \longrightarrow 2\ Cr^{3+} + 7\ H_2O \qquad \text{(reduction)}$$

Step 3:

$$2\ H_2O + H_3PO_2 \longrightarrow H_3PO_4 + 4\ H^+ \qquad \text{(oxidation)}$$
$$14\ H^+ + Cr_2O_7{}^{2-} \longrightarrow 2\ Cr^{3+} + 7\ H_2O \qquad \text{(reduction)}$$

Step 4:

$$2\ H_2O + H_3PO_2 \longrightarrow H_3PO_4 + 4\ H^+ + 4\ e^- \qquad \text{(oxidation)}$$
$$\quad 0 \qquad\quad 0 \qquad\qquad 0 \qquad +4 \qquad -4$$
$$\qquad\qquad 0 \quad = \quad 0$$

$$6\ e^- + 14\ H^+ + Cr_2O_7{}^{2-} \longrightarrow 2\ Cr^{3+} + 7\ H_2O \qquad \text{(reduction)}$$
$$-6 \qquad +14 \qquad -2 \qquad\qquad +6 \qquad 0$$

Step 5:

$$3[2\ H_2O + H_3PO_2 \longrightarrow H_3PO_4 + 4\ H^+ + 4\ e^-] \qquad \text{(oxidation)}$$
$$2[6\ e^- + 14\ H^+ + Cr_2O_7{}^{2-} \longrightarrow 2\ Cr^{3+} + 7\ H_2O] \qquad \text{(reduction)}$$

Step 6:

$$6\ H_2O + 3\ H_3PO_2 + 12\ \cancel{e} + 28\ H^+ + 2\ Cr_2O_7{}^{2-} \longrightarrow 3\ H_3PO_4 + 12\ H^+ + 12\ \cancel{e} + 4\ Cr^{3+} + 14\ H_2O$$

The equation has 6 $H_2O$ molecules on the left and 14 $H_2O$ molecules on the left. 6 $H_2O$ molecules are common species on both sides and are canceled in the following way:

$$6\ \cancel{H_2O} + 3\ H_3PO_2 + 28\ H^+ + 2\ Cr_2O_7{}^{2-} \longrightarrow 3\ H_3PO_4 + 12H^+ + 4\ Cr^{3+} + \overset{8}{\cancel{14}}\ H_2O$$

The same process is used for $H^+$:

$$3\ H_3PO_2 + \overset{16}{\cancel{28}}\ H^+ + 2\ Cr_2O_7{}^{2-} \longrightarrow 3\ H_3PO_4 + 12\ \cancel{H^+} + 4\ Cr^{3+} + 8\ H_2O$$

The final balanced equation,

$$3\ H_3PO_2{}_{(aq)} + 16\ H^+{}_{(aq)} + 2\ Cr_2O_7{}^{2-}{}_{(aq)} \longrightarrow 3\ H_3PO_4{}_{(aq)} + 4\ Cr^{3+}{}_{(aq)} + 8\ H_2O \qquad (1)$$

b.  $$MnO_4{}^-{}_{(aq)} + ClO_2{}^-{}_{(aq)} \longrightarrow MnO_2 + ClO_4{}^-{}_{(aq)} \qquad \text{(basic)}$$
$$+7-2 \qquad\quad +3-2 \qquad\qquad +4-2 \quad +7-2$$

Step 1:   (See oxidation numbers above).

$$ClO_2{}^- \longrightarrow ClO_4{}^- \qquad \text{(oxidation)}$$
$$+3 \qquad\quad +7$$

$$MnO_4{}^- \longrightarrow MnO_2 \qquad \text{(reduction)}$$
$$+7 \qquad\quad +4$$

Step 2:

$$4\ OH^- + ClO_2{}^- \longrightarrow ClO_4{}^- \qquad \text{(oxidation)}$$

(2 O's needed on left side, so 4 $OH^-$ added to left side.)

$$MnO_4{}^- \longrightarrow MnO_2 + 4\ OH^- \qquad \text{(reduction)}$$

(2 O's needed on right side, so 4 $OH^-$ added to right side.)

Step 3:
$$4\ OH^- + ClO_2^- \longrightarrow ClO_4^- + 2\ H_2O \qquad \text{(oxidation)}$$
$$2\ H_2O + MnO_4^- \longrightarrow MnO_2 + 4\ OH^- \qquad \text{(reduction)}$$

Step 4:
$$4\ OH^- + ClO_2^- \longrightarrow ClO_4^- + 2\ H_2O + 4\ e^- \qquad \text{(oxidation)}$$
$$\phantom{4\ }-4 \qquad -1 \qquad\qquad -1 \qquad\quad 0 \qquad\quad -4$$

$$3\ e^- + 2\ H_2O + MnO_4^- \longrightarrow MnO_2 + 4\ OH^- \qquad \text{(reduction)}$$
$$-3 \qquad\quad 0 \qquad -1 \qquad\qquad 0 \qquad\quad -4$$

Step 5:
$$3(4\ OH^- + ClO_2^- \longrightarrow ClO_4^- + 2\ H_2O + 4\ e^-)\ \text{(oxidation)}$$
$$4(3\ e^- + 2\ H_2O + MnO_4^- \longrightarrow MnO_2 + 4\ OH^-) \qquad \text{(reduction}$$

Step 6:
$$\cancel{12}\ OH^- + 3\ ClO_2^- + \cancel{12}\ e^- + \cancel{8}\ H_2O + 4\ MnO_4^- \longrightarrow 3\ ClO_4^- + \cancel{6}\ H_2O + \cancel{12}\ e^- + 4\ MnO_2 + \cancel{16}\ OH^-$$

$$3\ ClO_2^- + 2\ H_2O + 4\ MnO_4^- \longrightarrow 3\ ClO_4^- + 4\ MnO_2 + 4\ OH^-$$

---

## ANSWERS TO SELF-TEST

1. a
2. a
3. c
4. d
5. b
6. c
7. d
8. b
9. c
10. b

# RADIOACTIVITY AND NUCLEAR PROCESSES

# CHAPTER
# 17

## 17.1 Radioactivity

Radioactivity is the transformation of unstable atomic nuclei
into more stable nuclei.  Nuclei can become more stable by
emitting particles or energy or by capturing electrons from
outside the nucleus.  In the process, the nuclei change into
nuclei of another element.

Radioactive isotopes are called radioisotopes.  All isotopes
of elements beyond bismuth (atomic number 83) are radioactive.  Some elements with atomic numbers less than 84
also have radioisotopes.

Low atomic number nuclei have neutron-to-proton ratio, n/p,
very close to 1.  Atoms of atomic number greater than 83
have n/p ratios approaching 1.5.

Radioactive decay is the process in which the unstable
nucleus of a radioisotope forms a new nucleus.  Decay con-
tinues until a stable nucleus is achieved.

Three kinds of radioactive decay are named alpha, beta and
gamma decay (Table 17-2).

## 17.2 Alpha Decay

An alpha particle contains two protons and two neutrons.  It
is identical to a helium-4 nucleus with an atomic number of
2 and a mass number of 4.

$$\text{alpha particle} \quad {}_2^4\alpha$$

17.3   <u>Beta Decay</u>

There are three types of beta decay.

1.  Electron emission occurs when electrons are emitted from a nucleus.  Emitted electrons, called beta particles, are created at the instant of emission by conversion of a neutron to an electron and a proton.

$$\text{beta particle} \quad {}^{0}_{-1}\beta$$

2.  Positron emission occurs when a proton loses a small, positively charged particle, the positron, and becomes a neutron.

$$\text{positron} \quad {}^{0}_{1}\beta$$

3.  Electron capture occurs when an outside electron enters into the nucleus, changing a proton into a neutron.

$$\text{electron} \quad {}^{0}_{-1}e$$

17.4   <u>Gamma Decay</u>

Gamma rays are a highly penetrating form of energy that accompanies other types of decay.  Gamma rays have no mass or charge.

$$\text{gamma} \quad \gamma$$

17.5   <u>Nuclear Reactions</u>

In a nuclear reaction, the sum of the atomic numbers on the left side of the arrow is equal to the sum of the atomic numbers on the right side.  The same is true for mass numbers.

17.6   <u>Natural and Artificial Radioactivity</u>

Most of the naturally occurring radioisotopes decay very slowly and have existed since the formation of the earth. Natural isotopes with faster decay rates are continually formed by bombardment from secondary cosmic rays.

One of three radioactive disintegration series account for the decay of radioisotopes of atomic number greater than 81. The stable isotope of lead is the end product in each series.

Nuclear bombardment by alpha particles, neutrons or other subatomic particles, converts one element into another.

The transuranium elements (atomic number > 92) are synthesized by bombardment.

Transmutation is the conversion of one element into another. It can be accomplished by bombardment or natural radio-activity.

## 17.7 Detection and Measurement of Nuclear Radiation

Radiation emitted during nuclear reactions is called nuclear radiation. Some forms of nuclear radiation can cause ionization in the substances they strike. These types of radiation are called ionizing radiation (Table 17-3).

Nuclear radiation can be detected and measured by Geiger counters, scintillation counters, and film badges.

Different radioisotopes decay at different rates. The rate is measured by the length of time required for half of the atoms of a radioisotope to decay. This length of time is called the half-life (Figure 17-4).

Nuclear radiation measurements are reported in a variety of units depending on the purpose of the measurement. The activity of a sample measures the number of disintegrations in a given period of time. The dosage of a sample measures the amount of radiation delivered by the material. The energy level of the radiation measures the energy emitted by the sample (Table 17-5).

## 17.8 Radiation Safety

Nuclear radiation and X rays form ions and free radicals (particles with an unpaired electron) that react with nearby biological molecules in living tissue (Figure 17-5).

The more massive a particle, the less deeply it penetrates into tissue, and the more biological damage it can cause (Table 17-6).

The effects of different doses of nuclear radiation, measured in rems, are given in Table 17-7.

Exposure to nuclear radiation can be minimized by proper shielding, adequate distance from the source, and the use of low intensity sources.

## 17.9 Applications of Radiochemistry

X rays are high-energy ultraviolet light resulting from electron transitions within atoms.

The half-life of certain radioisotopes can be used for archeological dating.

Metabolic pathways and complex chemical systems can be studied using compounds tagged with radioisotopes.

Gamma rays and X rays are used in the diagnosis and treatment of numerous medical conditions.

## 17.10  Nuclear Energy

The Einstein equation indicates that matter and energy are interchangeable.

$$E = mc^2,$$

where E = energy
     m = mass
     c = speed of light

Since c is such a large number ($3 \times 10^{10}$ cm/sec), a very small amount of matter can release a large amount of energy.

The mass of a nucleus is lower than the mass of the protons and neutrons found in the nucleus.  This deficiency in mass is called the mass defect.  The mass defect arises when energy is released as particles combine to form a nucleus.

This energy, equivalent to the mass defect is called the nuclear binding energy.

The average binding energy is the nuclear binding energy of a particular nucleus divided by the number of protons and neutrons in the nucleus.  High binding energies indicate stability (Figure 17-9).

## 17.11  Nuclear Fission

In nuclear fission, a large nucleus breaks into two smaller nuclei.  Fission has to be initiated by neutron bombardment. Large amounts of energy are released due to the difference in the binding energies of the reactant nucleus and the product nuclei.

In a fission reaction, more neutrons can be produced than are consumed.

Critical mass is the smallest amount of fissionable isotope required for a self-sustaining nuclear chain reaction.

If the mass of the reactant is small, called a subcritical mass, the number of neutrons produced (that can cause further fission) averages less than one per reaction.  In this case, a chain reaction will not be sustained.

In a critical mass, the number of neutrons that can produce further fission averages one per reaction.  In this case, a chain reaction will be self-sustaining.

In a supercritical mass, more than one neutron (capable of causing further fission) is produced.  The chain reaction is explosive.

Controlled fission is carried out in nuclear reactors.

### 17.12 Nuclear Fusion

In nuclear fusion, two smaller atomic nuclei are combined to produce a heavier nucleus with greater stability (higher binding energy).

The fusion reactions that occur in the sun can be summarized as

$$^{1}_{1}H + {}^{2}_{1}H \longrightarrow {}^{3}_{2}He + energy$$

---

### SOLUTIONS TO SELECTED STUDY QUESTIONS AND PROBLEMS
### and
### PRACTICE PROBLEMS

16.  Write the complete equation for each nuclear reaction.

a.  Alpha decay by polonium -198.

All decay processes can be described by the following general equation:

$$isotope\ A \longrightarrow \left(\begin{array}{c}subatomic\ particles\\ and/or\ radiation\end{array}\right) + isotope\ B$$

There will be only one reactant, an isotope, and several products.

To write the $\alpha$-decay reaction for polonium-198, begin by writing the symbols of isotopes and particles given in the problem.  Look up the atomic number for polonium on the periodic table.

$$^{198}_{84}Po \longrightarrow {}^{4}_{2}\alpha + ?$$

Next find the atomic number of the product.  The sum of the reactant atomic numbers equals the sum of the product atomic numbers.

If Z is the atomic number of the unknown element, then

$$84 = 2 + Z$$
$$82 = Z$$

Look on the periodic table to find the element with atomic number 82.

$$^{198}_{84}Po \longrightarrow {}^{4}_{2}\alpha + {}^{?}_{82}Pb$$

Finally determine the mass number of lead in the same way as atomic number.

$$^{198}_{84}Po \longrightarrow {}^{4}_{2}\alpha + {}^{194}_{82}Pb$$

b.  Electron emission to form bismuth -210.

Electron emission is a decay reaction.  Since the electron is emitted from the nucleus, the symbol used is $^{0}_{-1}\beta$.

The atomic number of bismuth, given on the periodic table is 83.  Note that bismuth is formed, and is, therefore a product.

$$? \longrightarrow {}_{-1}^{0}\beta + {}_{83}^{210}\text{Bi}$$

The atomic number of the reactant is 82, which is lead

$$_{82}^{?}\text{Pb} \longrightarrow {}_{-1}^{0}\beta + {}_{83}^{210}\text{Bi}$$

The mass number doesn't change in the emission of a beta particle, (0 + 210 = 210).

$$_{82}^{210}\text{Pb} \longrightarrow {}_{-1}^{0}\beta + {}_{83}^{210}\text{Bi}$$

c.   Formation of osmium -188 from platinum -192.

First write the symbols with platinum as a reactant and osmium as a product.

$$_{78}^{192}\text{P+} \longrightarrow {}_{76}^{188}\text{Os}$$

The product isotope has a lower atomic number and mass number so a decay process occurred.

$$_{78}^{192}\text{P+} \longrightarrow {}_{76}^{188}\text{Os} + ?$$

Find the atomic number and mass number of the product.

$$_{78}^{192}\text{P+} \longrightarrow {}_{76}^{188}\text{Os} + {}_{2}^{4}?$$

The subscript and superscript indicate that the particle emitted is an alpha particle.

$$_{78}^{192}\text{P+} \longrightarrow {}_{76}^{188}\text{Os} + {}_{2}^{4}\alpha$$

Consult Appendix C in the text for the answer to part d.

***Practice Problem A

Complete the following nuclear equations:

a.   $_{6}^{14}\text{C} \longrightarrow {}_{-1}^{0}\beta + {}_{?}^{14}\text{N}$

b.   $_{92}^{238}\text{U} \longrightarrow {}_{2}^{4}\alpha + ?$

c. $\phantom{x}^{226}_{\phantom{2}88}\text{Ra} \longrightarrow ? + {}^{222}_{\phantom{2}86}\text{Rn}$

d. $\phantom{x}^{59}_{27}\text{Co} + {}^{1}_{0}\text{n} \longrightarrow$

e. $\phantom{x}^{27}_{13}\text{Al} + {}^{4}_{2}\alpha \longrightarrow {}^{1}_{0}\text{n} + ?$

47. Write the complete equation for each nuclear reaction.

a. Positron emission and gamma decay by cobalt −56.

First, write symbols for the isotope, particle and radiation given in the problem. Cobalt −56 is the only reactant species.

$$\phantom{x}^{56}_{27}\text{Co} \longrightarrow {}^{0}_{1}\beta + \gamma + ?$$

(Note that a positron has a symbol similar to a beta particle, or electron, except that the subscript is positive one.)

Next, find the atomic number of the unknown product.

$$27 = 1 + Z$$
$$26 = Z$$

Look on the periodic table to find the element with atomic number 26.

$$\phantom{x}^{56}_{27}\text{Co} \longrightarrow {}^{0}_{1}\beta + \gamma + {}^{?}_{26}\text{Fe}$$

Finally, determine the mass number of iron, in the same way as atomic number.

$$\phantom{x}^{56}_{27}\text{Co} \longrightarrow {}^{0}_{1}\beta + \gamma + {}^{56}_{26}\text{Fe}$$

d.  Formation of oxygen -18 by electron emission.

Write the symbol of the given isotope and particle.

$$? \longrightarrow {}^{18}_{?}O + {}^{0}_{-1}\beta$$

Look up the atomic number for oxygen on the periodic table.

$$? \longrightarrow {}^{18}_{8}O + {}^{0}_{-1}\beta$$

Determine the atomic number and mass number of the reactant isotope.

$${}^{18}_{7}? \longrightarrow {}^{18}_{8}O + {}^{0}_{-1}\beta$$

Look on the periodic table to find the element with atomic number 7.

$${}^{18}_{7}N \longrightarrow {}^{18}_{8}O + {}^{0}_{-1}\beta$$

Consult Appendix C in the text for answers to parts b, c, and e.

***Practice Problem B

Write the complete equation for each nuclear reaction.

a.  Transformation of californium -249 to curium -245

b.  Electron emission by berkelium -249.

c.  Formation of a proton by alpha particle bombardment of nitrogen -14.

---

SELF-TEST

Circle the correct answer in the following multiple choice questions.

1.  Which of the following is not a radioisotope?

    a   Th -215
    b.  Pb -206
    c.  Es -253
    d.  Po -218

2.  Which of the following symbols is correct for the given particle?

    a.  alpha particle, $^2_4\alpha$
    b.  positron, $^1_1H$
    c.  beta particle $^0_{-1}\beta$
    d.  gamma radiation $^1_0\gamma$

3.  Which of the following is not a mode of nuclear decay?

    a.  alpha emission
    b.  electron capture
    c.  neutron capture
    d.  positron emission

4.  The product of alpha particle emission by bismuth -210 is

    a.  thallium -206
    b.  gold -208
    c.  polonium -210
    dl  lead -209

5.  Which of the following statements is incorrect for radio-active disintegration series?

    a.  Many disintegrations may occur before a stable isotope is formed.
    b.  Some disintegrations involve the loss of a beta particle and some involve the loss of an alpha particle.
    c.  The end product of each radioactive disintegration series is a stable isotope of lead.
    d.  Only radioisotopes with atomic numbers greater than 91 (uranium and higher atomic number elements) decay by a disintegration series.

6.  Which one of the following statements is not true for ionizing radiation?

    a   Alpha radiation is low energy.
    b   Gamma rays have a greater penetrating power than do alpha and beta particles.

c.   The most penetrating forms of nuclear radiation cause the most biological damage.
d.   Ionizing radiation knocks electrons off of atoms in their path.

7.   Starting with a 4.0g sample of $^{234}_{90}Th$, how much remains at the end of 72 days?  The half-life of $^{234}_{90}Th$ is 24 days.

a.   2.0g
b.   1.0g
c.   .50g
d.   .25g

8.   A high average binding energy indicates

a.   a small mass defect.
b.   a stable nucleus.
c.   a large quantity of energy is required to hold the particles of the nucleus together.
d.   that the isotope is rarely found in nature.

9.   Which of the following reactions is a fission process?

a.   $^{235}_{92}U + {}^{1}_{0}n \longrightarrow {}^{139}_{54}Xe + {}^{95}_{38}Sr + 2\ {}^{1}_{0}n$

b.   $^{6}_{3}Li + {}^{1}_{1}H \longrightarrow {}^{4}_{2}He + {}^{3}_{2}He$

c.   $2\ {}^{2}_{1}H \longrightarrow {}^{4}_{2}He$

d.   $^{252}_{98}Cf + {}^{11}_{5}B \longrightarrow {}^{260}_{103}Lr + 2\ {}^{1}n$

10.   If a mass of fissionable material is such that each fission produces, on the average, more than one neutron that can cause further fission, then the mass of the fissionable material is said to be

a.   subcritical
b.   hypercritical
c.   supercritical
d.   critical

---

## ANSWERS TO PRACTICE PROBLEMS

Practice Problem A

a.   $^{14}_{6}C \longrightarrow {}^{0}_{-1}\beta + {}^{14}_{7}N$

b.   $^{238}_{92}U \longrightarrow {}^{4}_{2}\alpha + {}^{234}_{90}Th$

c.   $^{226}_{88}Ra \longrightarrow {}^{4}_{2}\alpha + {}^{222}_{86}Rn$

d.   $^{59}_{27}Co + {}^{1}_{0}n \longrightarrow {}^{60}_{27}Co$

e. $^{27}_{13}Al + ^{4}_{2}\alpha \longrightarrow ^{1}_{0}n + ^{30}_{15}P$

Reactions a, b, and c are decay reactions. Reactions d and e are bombardment reactions.

Practice Problem B

a. $^{249}_{98}Cf \longrightarrow ^{245}_{96}Cm + ^{4}_{2}\alpha$

b. $^{249}_{97}Bk \longrightarrow ^{0}_{-1}\beta + ^{249}_{98}Cf$

c. $^{14}_{7}N + ^{4}_{2}\alpha \longrightarrow ^{1}_{1}H + ^{17}_{8}O$

---

## ANSWERS TO SELF-TEST

1. b
2. c
3. c
4. a
5. d
6. c
7. c
8. b
9. a
10. c

# INTRODUCTION TO ORGANIC CHEMISTRY

# CHAPTER 18

---

SUMMARY OUTLINE

18.1    The Nature of Organic Chemistry

Organic chemistry is the study of most covalent carbon compounds.

Organic compounds have relatively low melting and boiling points, and are usually weak electrolytes or nonelectrolytes. The combustion of organic compounds produces carbon dioxide, water, heat, and light.

Carbon atoms form single, double, or triple covalent bonds to other carbon atoms and single bonds to atoms of hydrogen, nitrogen oxygen, phosphorus, sulfur and the halogens. Carbon atoms also form multiple bonds to atoms of nitrogen and oxygen.

18.2    Hydrocarbons

Binary molecular compounds containing only hydrogen and carbon are called hydrocarbons.  The carbon atoms are arranged in unbranched chains, in branched chains, in rings, or other more complicated structures.  Hydrogen atoms are bonded to the carbon atoms, to give each carbon atom four bonds.

Most hydrocarbons are colorless, nonpolar molecules which are insoluble in water, but soluble in one another, and less dense than water.

The physical state of hydrocarbons at room temperature depends on their formula weight:

|  |  |
|---|---|
| Gases, | less than 5 carbons |
| Liquids, | 5-18 carbons |
| Solids, | more than 18 carbons |

Hydrocarbons are classified as aliphatic or aromatic. Aliphatic hydrocarbons, usually unbranched or branched chains, are divided into three major subclasses: alkanes, alkenes, and alkynes.

## 18.3  Alkanes

Alkanes contain only single bonds. Alkanes are known as saturated hydrocarbons because each carbon atom is bonded to four other atoms.

All open-chain alkanes have the general formula $C_nH_{2n+2}$, where n is the number of carbon atoms. The names and formulas of the first ten linear (unbranched) alkanes are given below:

| | | | |
|---|---|---|---|
| $CH_4$ | methane | $C_6H_{14}$ | hexane |
| $C_2H_6$ | ethane | $C_7H_{16}$ | heptane |
| $C_3H_8$ | propane | $C_8H_{18}$ | octane |
| $C_4H_{10}$ | butane | $C_9H_{20}$ | nonane |
| $C_5H_{12}$ | pentane | $C_{10}H_{22}$ | decane |

When there are four single bonds to a carbon atom, the bond angles will be 109.5° (Figure 18-1). There is free rotation about single bonds resulting in various shapes for each molecule. Linear molecules are not actually linear; the carbon atoms have a zigzag arrangement.

Three types of formulas are used to represent organic molecules on paper.

1. Molecular formulas show number and kind of atoms in a molecule.

   ethane  $C_2H_6$

2. Condensed structural formulas give major structural features of a molecule. Each carbon is shown, with elements bonded to it written to the right.

   ethane  $CH_3CH_3$ or $CH_3-CH_3$

3. Structural formulas show major structural features including bonds.

   ethane
   $$H-\overset{\displaystyle H}{\underset{\displaystyle H}{\overset{|}{\underset{|}{C}}}}-\overset{\displaystyle H}{\underset{\displaystyle H}{\overset{|}{\underset{|}{C}}}}-H$$

Compounds having the same molecular formula but different arrangements of atoms are called iosmers. Alkanes with four or more carbons have isomers (Figure 18-2). The physical properties of isomers differ.

The higher the number of carbon atoms the more isomers an alkane will have (Table 18-2).

18.4  <u>Naming Alkanes</u>

Organic compounds are named by a system proposed by the International Union of Pure and Applied Chemistry.  The following are IUPAC rules for naming open-chain alkanes.

1.  Use the ending -<u>ane</u> for all alkane names.
2.  Attach a prefix to this ending to indicate the number of carbon atoms in the longest continuous chain (parent chain) of carbon atoms.  The prefixes are

$$
\begin{array}{llll}
\text{meth} & - \ 1C & \text{hex} & \ \ \ 6C \\
\text{eth} & - \ 2C & \text{hept} & - \ 7C \\
\text{prop} & - \ 3C & \text{oct} & - \ 8C \\
\text{but} & - \ 4C & \text{non} & - \ 9C \\
\text{pent} & - \ 5C & \text{dec} & - \ 10C \\
\end{array}
$$

3.  Assign a number to each carbon in the parent chain beginning at the end nearest the branch.

$$
\begin{array}{ccccc}
1 & 2 & 3 & 4 & 5 \\
\end{array}
$$
$$
CH_3-CH_2-CH-CH-CH-CH_2-CH_3
$$
$$
\underset{\displaystyle CH_3\ CH_3\ \underset{\displaystyle \underset{\displaystyle CH_2 \ 7}{|}}{CH_2}\ 6}{|\ \ \ |\ \ \ |}
$$
$$
CH_3 \ 8
$$

4.  Hydrocarbon branches having only single bonds are called alkyl groups.  Each alkyl group is an alkane molecule minus one hydrogen.  Change the alkane name to alkyl.

$$
\begin{array}{ll}
CH_3CH_3 & \text{ethane} \\
CH_3CH_2- & \text{ethyl} \\
\end{array}
$$

5.  Use the name of the alkyl group as a prefix to the name of the parent alkane.  Alkyl groups are written in alphabetical order.  Place the number locating the alkyl group in front of the alkyl name.  Identical alkyl groups are written with Greek prefixes di-, tri-, tetra-, penta-, and so on, to indicate the number present in the molecule.  Commas separate numbers, and hyphens separate numbers from words.

5-ethyl-3,4-dimethyloctane

## 18.5 Alkenes and Alkynes

Unsaturated hydrocarbons have one or more multiple bonds. With multiple bonds fewer hydrogen atoms are found in the molecule. The molecule is "less saturated" with hydrogen atoms than an alkane with the same number of carbon atoms.

Alkenes are aliphatic hydrocarbons having one or more double bonds between carbon atoms.

Open-chain alkenes with one double bond have the general formula, $C_nH_{2n}$ (Table 18-3).

The bond angles around the carbon atoms connected by a double bond are 120° (Figure 18-3).

Groups connected by a double bond cannot rotate about the double bond. Because of this, alkenes can have an additional type of isomerism called geometric isomerism. Alkenes having two different groups on each of the double bonded carbon atoms can have two different geometries.

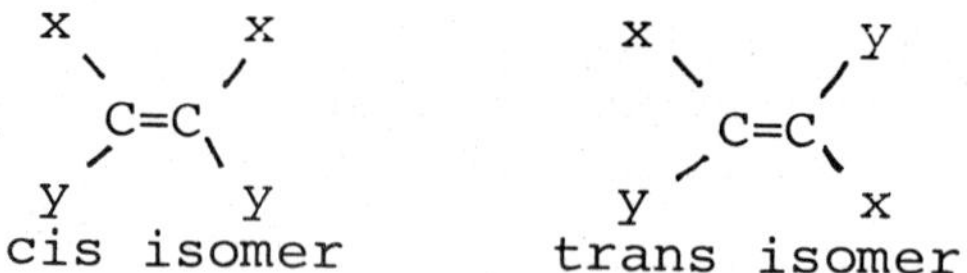

Alkynes are aliphatic hydrocarbons having one or more triple bonds. Alkynes contain fewer hydrogens per carbon atom than alkenes, so they are less saturated than the alkenes.

Open-chain alkynes with one triple bond have the general formula $C_nH_{2n-2}$ (Table 18-4).

The bond angles around the carbon atoms connected by the triple bond are 180° (Figure 18-5).

## 18.6 Naming Alkenes and Alkynes

The following are IUPAC rules for naming open-chain alkenes.

1. Use the ending -ene for all alkene names.
2. Attach a prefix to this ending to indicate the number of carbon atoms in the longest continuous chain (parent chain) of carbon atoms containing the double bond. (See Naming Alkanes, rule 2 for a list of the prefixes.)
3. Assign a number to each carbon in the parent chain giving the first carbon of the double bond the lowest possible number.

4.  Groups attached to the parent chain are identified individually by number and names, as with alkanes.

$$\overset{6}{CH_3}-\overset{5}{CH_2}-\overset{4}{CH_2}-\overset{3}{C}\!-\!-\!CH_2CH_2CH_3$$

with side chain: $CH_2$ (2), $CH_3$ (1)

3-propyl-2-hexene

5.  If the alkene contains more than one double bond, the ending is changed to -<u>diene</u>, -<u>triene</u>, and so on to indicate the number of double bonds.  The location of each double bond is identified by number.

Alkynes are named in the same, except that the ending for alkynes is -<u>yne</u>.

## 18.7    <u>Cyclic Aliphatic Hydrocarbons</u>

The carbon atoms of aliphatic hydrocarbons can be joined in rings.  These cyclic compounds are named by adding the prefix <u>cyclo-</u> to the name of the corresponding open-chain alkane, alkene, or alkyne.

Cyclic molecules can be drawn using geometric figures similar to the one shown below,

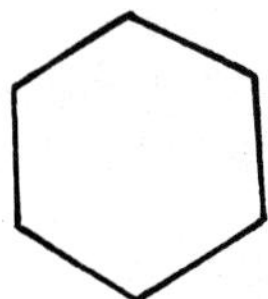

cyclohexane

where each line represents a C–C bond and each corner is a carbon atom with the appropriate number of hydrogen atoms to satisfy carbon's valence of 4 (Figure 18-6).

## 18.8    <u>Benzene and Aromatic Hydrocarbons</u>

Aromatic hydrocarbons are compounds structurally related to benzene, $C_6H_6$.

Two structures for benzene were proposed in 1865 by Kekule:

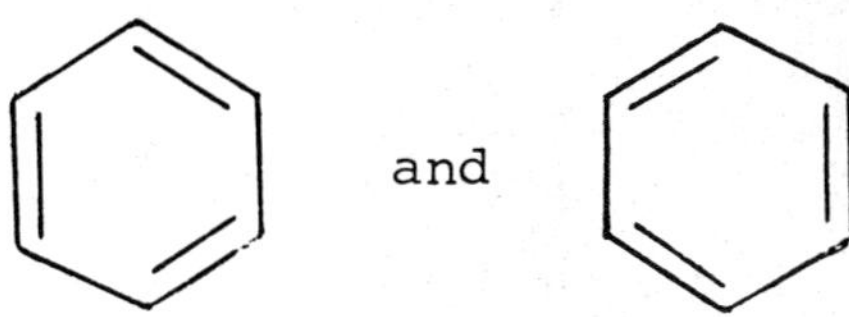

and

The three double bonds are actually spread over all carbon atoms in the ring.  A better representation of the molecule is shown below.

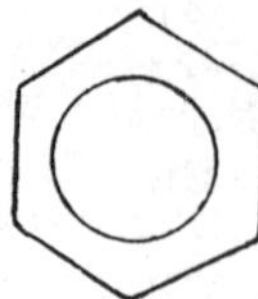

Aromatic hydrocarbons have one or more double bonds spread over (delocalized) all carbon atoms in the ring.  This imparts unusual stability to the molecule.

Benzene derivatives are formed when one or more hydrogen atoms on the ring are replaced by other atoms or groups (Table 18-5).

Polycyclic aromatic hydrocarbons consist of molecules containing two or more aromatic rings fused together.

18.9    Classification of Organic Compounds by Functional Group

A functional group is a distinctive group of atoms attached to the inert alkyl group support structure.  Functional groups always react in the same way.  Functional groups are used to classify and recognize types of organic compounds.  Table 18-6 lists the common functional groups.

Organic molecules can contain more than one functional group.  The chemical behavior of these molecules will be a composite of the individual behaviors of each functional group.

18.10    Orbital Hybridization in Hydrocarbon Molecules (optional)

The formation of a covalent bond corresponds to the overlap of two atomic orbitals.  The new electron cloud, formed by the merged atomic orbitals, is called a molecular orbital.  A molecular orbital can contain a maximum of two electrons (Figure 18-7).

A single covalent bond is called a sigma bond because the symmetrical molecular orbital is along the axis joining the nuclei of the two bonded atoms (Figure 18-8).

When a carbon atom bonds to four atoms, the atomic orbitals of carbon's valence shell form four hybrid orbitals called $sp^3$ orbitals.  The four $sp^3$ orbitals point toward the corners of a regular tetrahedron, with angles of 109.5° between the orbitals (Figures 18-9, 10, 11).  The $sp^3$ orbitals overlap with atomic orbitals of other atoms to form four sigma bonds (Figure 18-12).

When a carbon atom bonds to three atoms (in this case there is a double bond on carbon) only three hybrid orbitals are needed.  The s orbital and two p orbitals of carbon's valence shell form three hybrid orbitals called $sp^2$ orbitals.  The three $sp^2$ orbitals point toward the corners of an equilateral triangle with angles of 120° between orbitals (Figures 18-13, 14a).  The remaining p orbital is perpendicular to the plane of the three $sp^2$ orbitals (Figure 18-14b).

The $sp^2$ orbitals overlap with atomic orbitals of other atoms to form three sigma bonds.  The unhybridized p orbital can overlap side-to-side with a parallel p orbital from one of the carbon atoms sigma-bonded to an $sp^2$ orbital (Figure 18-15). The molecular orbital formed from side-to-side overlap of parallel p orbitals is called a pi molecular orbital.  A pi molecular orbital corresponds to a second bond, a pi bond between the two carbon atoms.  Thus, a double bond is composed of a sigma bond and a pi bond.

In benzene, sigma bonds are formed by overlapping $sp^2$ orbitals.  The six p orbitals can overlap above and below the ring, allowing delocalization of p electrons among all carbon atoms (Figure 18-16).

When a carbon atom bonds to two other atoms (in this case there is a triple bond on carbon) only two hybridized orbitals are needed.  The s orbital and one p orbital of carbon's valence shell form two hybridized orbitals called sp orbitals. The two sp orbitals lie in a straight line pointing in opposite directions, with an angle of 180° between them (Figure 18-17, 18a).  Two remaining p orbitals that are not hybridized are perpendicular to each other and the axes of the sp orbitals (Figure 18-18b).

The sp orbitals overlap with atomic orbitals of other atoms to form two sigma bonds.  The remaining p orbitals can overlap side-to-side with parallel p orbitals of another carbon atom to form two pi bonds (Figure 18-19).  A triple bond is composed of a sigma bond and two pi bonds.

---

### SOLUTIONS TO SELECTED STUDY QUESTIONS AND PROBLEMS
### and
### PRACTICE PROBLEMS

---

13.     Draw structural formulas for the three isomers of pentane and derive the IUPAC name of each.

First, write the molecular formula for pentane, an open-chain, five-carbon alkane.

$$C_nH_{2n+2} = C_5H_{2(5)+2}$$

$$= C_5H_{12}$$

The following guidelines can be used to draw isomers:

   1.   Begin with a straight-chain structure.  (It is helpful to draw only the carbon skeleton until all isomers are

found.  Then add hydrogens to complete the four bonds on carbon.)

$$C-C-C-C-C$$

2.  Draw a branched structure for the second isomer beginning with a straight-chain with one less carbon.

$$C-C-C-C$$

Use the other carbon as a branch.  Place it in as many different positions as possible.

$$
\begin{array}{ccc}
C-\underset{|}{C}-C-C & & C-C-\underset{|}{C}-C \\
C & \underset{|}{C} & C \\
& C-\overset{|}{C}-C-C &
\end{array}
$$

Note that there is only one branched isomer for a 4-C parent chain.  (All three isomers have the methyl group on the second carbon.)

3.  Draw additional isomers with parent chains of increasingly smaller length.  There can now be several branches with one carbon, or one branch with several carbons.  Locate the branches in as many different positions as possible.

$$
\begin{array}{c}
C \\
| \\
C-C-C \\
| \\
C
\end{array}
$$

A two carbon branch (ethyl) doesn't result in a new isomer for pentane.  The branch connects with the 3-C parent chain to make a longer 4-C chain with a methyl group on the second carbon.  That isomer has already been drawn (see isomer given in step 2).

$$
\begin{array}{c}
^1C-^2C-C \\
3\ \boxed{\overset{|}{\underset{|}{C}}} \\
4\ \boxed{C} \quad \text{ethyl}
\end{array}
$$

4.  Draw in all hydrogens on the correct skeletons.  Check to make sure the number of hydrogens is consistent with the molecular formula and each carbon has four bonds.  The condensed structural formulas for the three isomers of $C_5H_{12}$ are:

$$CH_3-CH_2-CH_2-CH_3 \qquad CH_3-\underset{\underset{CH_3}{|}}{CH}-CH_2-CH_3$$

$$CH_3-\underset{\underset{CH_3}{|}}{\overset{\overset{CH_3}{|}}{C}}-CH_3$$

The IUPAC name for each isomer can be found using the guidelines given in section 18.4.

$CH_3-CH_2-CH_2-CH_2-CH_3$ pentane

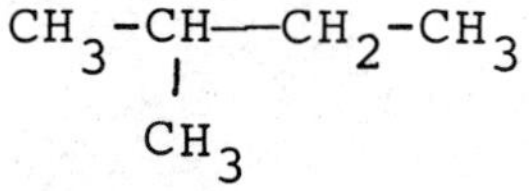

2-methylbutane

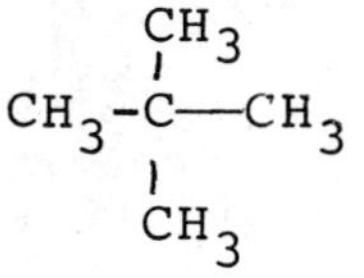

dimethylpropane (There is only one possible position for the methyl groups, so numbers are unnecessary.)

14. Draw the structural formula of each alkane.

a. 2-methylpentane

To draw structural formulas, begin by drawing the skeleton (carbons only) for the parent-chain. A pentane parent chain has 5 carbons.

C-C-C-C-C

Next, add the branches. A methyl group, $-CH_3$ is located on the second carbon.

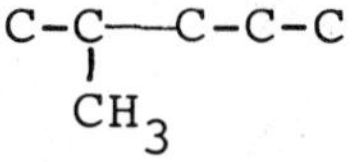

Finish the structure by adding enough hydrogens to give each carbon 4 bonds. The condensed structural formula is

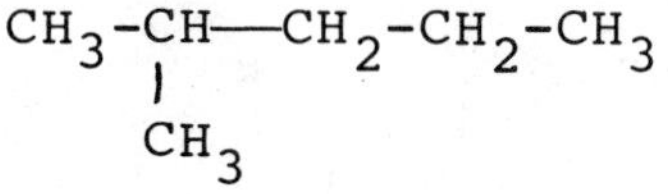

d. 3,4-diethyl-5-methylnonane

Nonane is a nine carbon parent chain.

C-C-C-C-C-C-C-C-C

Draw ethyl groups, $-CH_2CH_3$ on the third and fourth carbons.

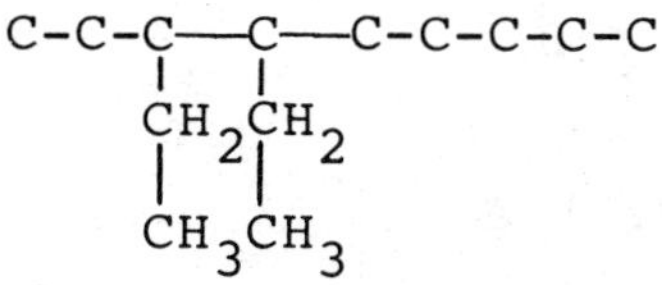

Draw a methyl group, $-CH_3$, on the fifth carbon.

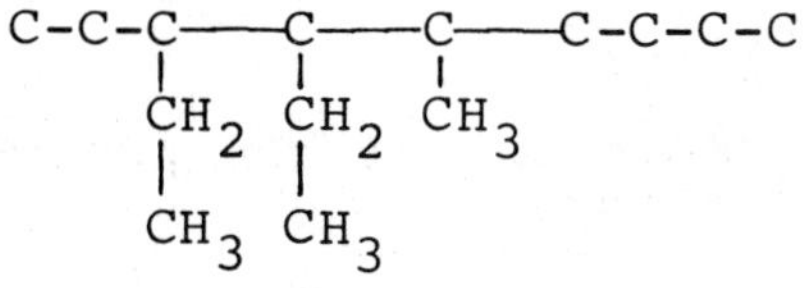

Add hydrogens to complete the molecule.  The condensed structural formula is

$$CH_3-CH_2-CH-CH-CH-CH_2-CH_2-CH_2-CH_3$$

with substituents:

$$CH_2 \quad CH_2 \quad CH_3$$
$$CH_3 \quad CH_3$$

Consult Appendix C in the text for answers to parts b, c, and e.

19.  Designate each alkene as <u>cis</u>, <u>trans</u>, or neither.

a.

$$\begin{array}{ccc} CH_3 & & CH_3 \\ & C=C & \\ H & & H \end{array}$$

The two different groups on carbon are arranged so that the hydrogen atoms lie on the same side of the double bond.  This molecule is <u>cis</u>- 2-butane.

b.

$$\begin{array}{ccc} CH_3-CH_2-CH_2 & & H \\ & C=C & \\ H & & H \end{array}$$

This molecule does not exhibit cis-trans isomerism because the first carbon doesn't have two different groups on it.

c.

$$\begin{array}{ccc} CH_3-CH_2 & & H \\ & C=C & \\ H & & CH_3 \end{array}$$

The hydrogen atoms lie on opposite sides of the double bond.  This molecule is <u>trans</u>-2-pentene.

d.

$$\begin{array}{ccc} CH_3 & & CH_2-CH_3 \\ & C=C & \\ H & & H \end{array}$$

The hydrogen atoms lie on the same side of the double bond.  This molecule is <u>cis</u>-2-pentene.

49.  Draw the structural formula of each of the following.

a.  3-heptene

Draw the carbon skeleton for the parent chain.  Heptene has seven carbon atoms.

$$C-C-C-C-C-C-C$$

For unsaturated hydrocarbons, locate the multiple bonds next.  The double bond is found between the third and fourth carbon.  The number 3 indicates the position of the double bond.

$$C-C-C=C-C-C-C$$

Complete the structure by adding hydrogens to give four bonds to each carbon.

$$CH_3-CH_2-CH=CH-CH_2-CH_2-CH_3$$

b.    2-methyl-3-octyne

Draw an 8 carbon skeleton.

$$C-C-C-C-C-C-C-C$$

Locate the triple bond between the third and fourth carbon.

$$C-C-C\equiv C-C-C-C-C$$

Add the methyl branch on the second carbon.

$$\begin{array}{c} C-C \!-\!\!-\! C\equiv C-C-C-C-C \\ \quad | \\ \quad CH_3 \end{array}$$

Complete the structure by adding hydrogens to give four bonds to each carbon.  The condensed structural formula for 2-methyl-3-octyne is

$$\begin{array}{c} CH_3-CH\!-\!\!-\! C\equiv C-CH_2-CH_2-CH_2-CH_3 \\ \quad\quad | \\ \quad\quad CH_3 \end{array}$$

g.    2,4,6-decatriene

Draw a 10 carbon skeleton

$$C-C-C-C-C-C-C-C-C-C$$

There are three double bonds in this molecule (-triene) located after the second carbon, the fourth carbon and the sixth carbon.  Number the carbon atoms to avoid confusion.

$$\begin{array}{c} 1\ 2\ 3\ 4\ 5\ 6\ 7\ 8\ 9\ 10 \\ C-C=C-C=C-C=C-C-C-C \end{array}$$

Complete the structure by adding hydrogens to give four bonds to each carbon.

$$CH_3-CH=CH-CH=CH-CH=CH-CH_2-CH_2-CH_3$$

Consult Appendix C in the text for answers to parts c, d, e, and f.

***Practice Problem A

Draw the structural formula of each aliphatic hydrocarbon.

a.  3.3-dimethyl-1-pentene

b.  5-ethyl-3-octyne

c.  1,3-pentadiene

d.  2,2,3-trimethylbutane

e.  <u>trans</u>-2-hexene

***Practice Problem B

Derive the IUPAC name of each of the following aliphatic hydrocarbons.

a.

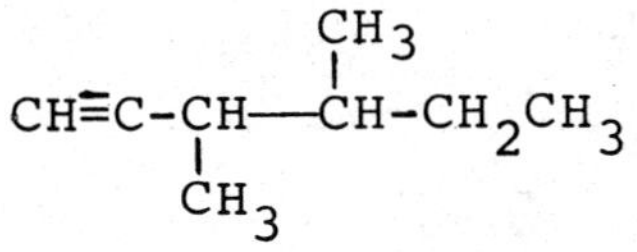

$$CH{\equiv}C\text{-}CH\text{---}\underset{\displaystyle CH_3}{\overset{\displaystyle CH_3}{\underset{|}{\overset{|}{CH}}}}\text{-}CH_2CH_3$$

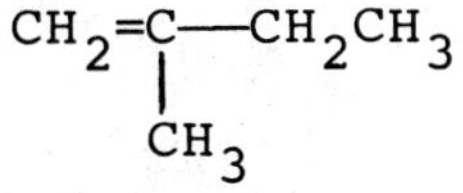

b.  $CH_2{=}\underset{\displaystyle CH_3}{\overset{|}{C}}\text{---}CH_2CH_3$

c.  $CH_3\text{-}\underset{\displaystyle CH_3}{\overset{|}{CH}}\text{---}CH{=}\underset{\displaystyle CH_3}{\overset{|}{C}}\text{---}CH_3$

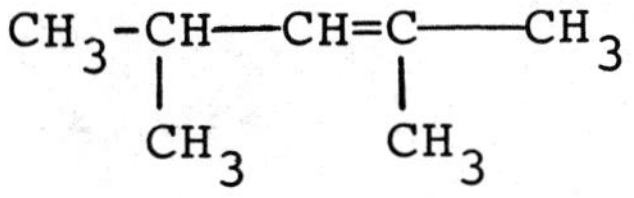

d.

---

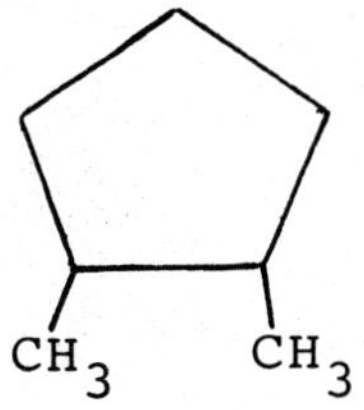

SELF-TEST

===

Circle the correct answer in the following multiple choice questions.

1.    All of the following are alkanes except

    a.  $C_4H_{10}$                          c.  $CH_3\text{-}CH_2\text{-}CH_2\text{-}CH_3$
    b.  2-methylpentane              d.  $C_6H_{12}$

2.  Which of the following statements is incorrect for alkanes?

a.  The bond angles about each carbon are 120°.
c.  Linear alkanes are actually zigzag.
d.  Each carbon is bonded to four other atoms.
e.  Alkanes with four or more carbons have isomers.

3.  The correct IUPAC name for the following molecule is

$$CH_3CH_2CHCH_2CH=C-CH_3$$

with $CH_3$ on the third carbon and $CH_2CH_3$ on the sixth carbon.

a.  2-ethyl-5-methyl-2-heptene
b.  3,6-dimethyl-3-octene
c.  6-ethyl-3-methyl-5-heptene
d.  3,6-dimethyl-6-octene

4.  Which one of the following is not an isomer of hexane?

a.  2,2-dimethylbutane
b.  3-methylpentane
c.  2,3-dimethylbutane
d.  2-ethylpentane

5.  Which of the following molecules is the least saturated?

a.  $CH_3-CH_2-CH_2CH_3$
b.  $CH_2=CH=CH-CH_3$
c.  $CH_3-CH=CH-CH_3$
d.  $CH_3-C\equiv C-CH_3$

6.  The molecular formula for the following is

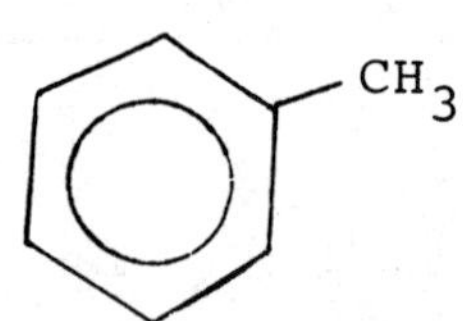

a.  $C_6H_8$
b.  $C_6H_9$
c.  $C_7H_8$
d.  $C_7H_9$

7.  Which of the following functional groups is correctly identified?

a.  $CH_3OH$,      aldehyde

b.  $CH_3\overset{\text{O}}{\overset{\|}{C}}-OH$,      carboxylic acid

$$\text{c.} \quad CH_3\overset{\overset{\displaystyle O}{\|}}{C}-H, \qquad \text{ketone}$$

$$\text{d.} \quad CH_3-\overset{\overset{\displaystyle O}{\|}}{C}-CH_3, \qquad \text{ester}$$

8. How many sigma bonds and pi bonds are found on the third carbon in

$$CH_3-CH_2-CH=CH-CH_2CH_2-CH_3$$

   a.  3 sigma bonds, 1 pi bond
   b.  2 sigma bonds, 2 pi bonds
   c.  4 sigma bonds, 0 pi bonds
   d.  1 sigma bond,  1 pi bond

9. Which of the following statements is incorrect for $sp^2$ hybrid orbitals?

   a.  The orbitals lie in a plane.
   b.  The angles between orbitals are 120°.
   c.  An unhydridized p orbital is perpendicular to the plane of the $sp^2$ orbitals.
   d.  A carbon with $sp^2$ orbitals will be sigma bonded to two other atoms.

10. The following molecule that is not an aliphatic hydrocarbon is

   a.  hexane
   b.  cyclohexane
   c.  2,4-hexadiene
   d.  methylbenzene

---

## ANSWERS TO PRACTICE PROBLEMS

Practice Problem A

$$\text{a.} \quad CH_3-CH_2-\overset{\overset{\displaystyle CH_3}{|}}{\underset{\underset{\displaystyle CH_3}{|}}{C}}-CH=CH_2$$

$$\text{b.} \quad CH_3-CH_2C\equiv C-\overset{|}{\underset{\underset{\underset{\underset{\displaystyle CH_3}{|}}{CH_2}}{|}}{CH}}-CH_2-CH_2-CH_3$$

$$\text{c.} \quad CH_2=CH-CH=CH-CH_3$$

$$\text{d.} \quad CH_3-\overset{\overset{\displaystyle CH_3}{|}}{\underset{\underset{\displaystyle CH_3}{|}}{C}}-\overset{}{\underset{\underset{\displaystyle CH_3}{|}}{CH}}-CH_3$$

e.

$$CH_3-CH_2-CH_2 \quad\quad H$$
$$C=C$$
$$H \quad\quad CH_3$$

The condensed structural formula,

$$CH_3CH_2CH_2-CH=CH-CH_3$$

does not distinguish the <u>cis</u>- isomer from the <u>trans</u>-isomer.  The groups on the carbon with the double bond must be clearly shown.

Practice Problem B

a.  3.4-dimethyl-1-hexyne
b.  2-methyl-1-butene
c.  2,4-dimethyl-2-pentene
d.  1,2-dimethylcyclopentane
    (number the branches on the ring with the lowest numbers possible, beginning on any carbon in the ring.)

---

## ANSWERS TO SELF-TEST

1.  d
2.  a
3.  b
4.  d
5.  d
6.  c
7.  b
8.  a
9.  d
10.  d